CISCO

D1997211

CCNP ROUTE
642-902
Quick Reference

ciscopress.com

Denise Donohue, CCIE No. 9566

CCNP
ROUTE 642-902 Quick Reference

Denise Donohue, CCIE No. 9566

Cisco Press

800 East 96th Street

Indianapolis, Indiana 46240 USA

CCNP ROUTE 642-902 Quick Reference

Denise Donohue

Published by:
Cisco Press
800 East 96th Street
Indianapolis, IN 46240 USA

Printed in the United States of America

Third Printing: March 2012

ISBN-13: 978-1-58720-317-6

ISBN-10: 1-58720-317-0

Publisher
Paul Boger

Associate Publisher
Dave Dusthimer

Executive Editor
Brett Bartow

Project Editor
Jennifer Gallant

Managing Editor
Patrick Kanouse

Editorial Assistant
Vanessa Evans

Designer
Sandra Schroeder

Warning and Disclaimer

This book is designed to provide information about networking. Every effort has been made to make this book as complete and as accurate as possible, but no warranty or fitness is implied.

The information is provided on an "as is" basis. The authors, Cisco Press, and Cisco Systems, Inc. shall have neither liability nor responsibility to any person or entity with respect to any loss or damages arising from the information contained in this book or from the use of the discs or programs that may accompany it.

The opinions expressed in this book belong to the author and are not necessarily those of Cisco Systems, Inc.

Trademark Acknowledgments

All terms mentioned in this book that are known to be trademarks or service marks have been appropriately capitalized. Cisco Press or Cisco Systems, Inc., cannot attest to the accuracy of this information. Use of a term in this book should not be regarded as affecting the validity of any trademark or service mark.

Corporate and Government Sales

The publisher offers excellent discounts on this book when ordered in quantity for bulk purchases or special sales, which may include electronic versions and/or custom covers and content particular to your business, training goals, marketing focus, and branding interests. For more information, please contact:

U.S. Corporate and Government Sales 1-800-382-3419
corpsales@pearsontechgroup.com

For sales outside the United States please contact:

International Sales international@pearsoned.com

Feedback Information

At Cisco Press, our goal is to create in-depth technical books of the highest quality and value. Each book is crafted with care and precision, undergoing rigorous development that involves the unique expertise of members from the professional technical community.

Readers' feedback is a natural continuation of this process. If you have any comments regarding how we could improve the quality of this book, or otherwise alter it to better suit your needs, you can contact us through email at feedback@ciscopress.com. Please make sure to include the book title and ISBN in your message.

We greatly appreciate your assistance.

Americas Headquarters
Cisco Systems, Inc.
170 West Tasman Drive
San Jose, CA 95134-1706
USA
www.cisco.com
Tel: 408 526-4000
800 553-NETS (6387)
Fax: 408 527-0883

Asia Pacific Headquarters
Cisco Systems, Inc.
168 Robinson Road
#28-01 Capital Tower
Singapore 068912
www.cisco.com
Tel: +65 6317 7777
Fax: +65 6317 7799

Europe Headquarters
Cisco Systems International BV
Haarlerbergpark
Haarlerbergweg 13-19
1101 CH Amsterdam
The Netherlands
www-europe.cisco.com
Tel: +31 0 800 020 0791
Fax: +31 0 20 357 1100

Cisco has more than 200 offices worldwide. Addresses, phone numbers, and fax numbers are listed on the Cisco Website at **www.cisco.com/go/offices.**

About the Authors

Denise Donohue, CCIE No. 9566, is a senior solutions architect for ePlus Technology, a Cisco Gold partner. She works as a consulting engineer, designing networks for ePlus' customers. Prior to this role, she was a systems engineer for the data consulting arm of SBC/AT&T. She has co-authored several Cisco Press books in the areas of route/switch and voice. Denise has worked as a Cisco instructor and course director for Global Knowledge and was a network consultant for many years. Her areas of specialization include route/switch, voice, and data center.

About the Technical Editors

'Rhette (Margaret) Marsh has been working in the networking and security industry for over ten years, and has extensive experience with internetwork design, IPv6, forensics, and greyhat work. She currently is a design consultant for Cisco in San Jose, CA, and works primarily with the Department of Defense and contractors. Before this, she worked extensively both in the financial industry as a routing and switching and design/security consultant and also in an attack attribution and forensics context. She currently holds a CCIE in Routing and Switching (CCIE No. 17476), CCNP, CCDP, CCNA, CCDA, CISSP and is working towards her Security and Design CCIEs. In her copious free time, she enjoys number theory, arcane literature, cycling, hiking in the redwoods, sea kayaking, and her mellow cat, lexx.

Contents at a Glance

Contents

Icons Used in This Book

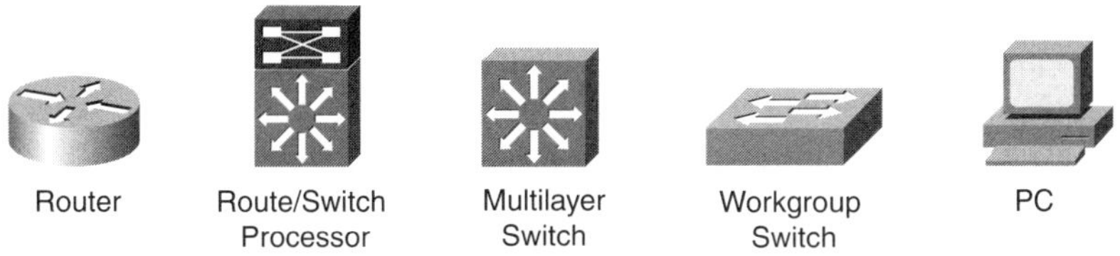

Command Syntax Conventions

The conventions used to present command syntax in this book are the same conventions used in the IOS Command Reference. The Command Reference describes these conventions as follows:

- **Boldface** indicates commands and keywords that are entered literally as shown. In actual configuration examples and output (not general command syntax), boldface indicates commands that are manually input by the user (such as a **show** command).
- *Italic* indicates arguments for which you supply actual values.
- Vertical bars (|) separate alternative, mutually exclusive elements.
- Square brackets ([ ]) indicate an optional element.
- Braces ({ }) indicate a required choice.
- Braces within brackets ([{ }]) indicate a required choice within an optional element.

Introduction

The Cisco Certified Network Professional (CCNP®) Routing and Switching certification validates knowledge and skills required to install, configure and troubleshoot converged local and wide area networks with 100 to 500 or more nodes. With a CCNP Routing and Switching certification, a network professional demonstrates the knowledge and skills required to manage the routers and switches that form the network core, as well as edge applications that integrate voice, wireless, and security into the network.

The *CCNP ROUTE 642-902 Quick Reference* was written to help you prepare for the *ROUTE 642-902* exam in the CCNP Routing and Switching certification. Some readers tell us that they use this book before beginning their exam preparation, to find which areas they are weak in. This helps target their studying. Others use it after studying or taking the course as a concise learning resource during their final preparation for the exam.

This book will also help once your exams are over, when you need a quick answer about a technology, or a reminder about configuration steps.

Who Should Read This Book?

Current and aspiring network engineers will find this book useful in two ways. First, those preparing for the CCNP Routing and Switching certification will appreciate the targeted review of exam topics. It will help them understand the technologies, not just memorize questions. This will lead to success on the exam and improved on-the-job performance. Secondly, the book serves as a reference for those not pursuing the certification. Its short descriptions and many examples come in handy when you need a fast answer to a question, or to configure something quickly. It deserves a place on every network engineer's bookshelf.

How This Book Is Organized

Following is a description for each section.

Chapter 1: Planning for Complex Networks

This chapter describes some different design models in use, provides an overview of routing protocols, and introduces methods of implementation planning.

Chapter 2: EIGRP

This chapter provides an in-depth description of EIGRP operation and configuration, including neighbor establishment and route exchange. It covers using EIGRP with Frame Relay, Ethernet over MPLS (EoMPLS) and Layer 3 MPLS VPNs. It also includes planning an EIGRP implementation, ways to optimize EIGRP, and securing EIGRP through authentication.

Chapter 3: OSPF

Chapter 3 describes OSPF's structure and operation. It covers OSPF design requirements, neighbor establishment, and LSA information. The configuration portion provides implementation planning, as well as OSPF configuration for LANs and WANs (including MPLS WANs.) The chapter additionally covers optimizing and securing OSPF.

Chapter 4: Optimizing Routing

This chapter examines various methods of controlling routing updates, such as route maps, prefix lists, and distribute lists. It describes how to configure route maps, and how to use them for policy-based routing, controlling route redistribution, and tagging routes. Route redistribution planning and isses are described, redistribution configuration is demonstrated.

Chapter 5: Path Control

Chapter 5 covers some additional ways to control the path that traffic takes. These include offset-lists, IP SLA, policy-based routing, Optimized Edge Routing, and VRFs. IP SLA tracking is shown along with route maps to provide backup routes when using policy-based routing.

Chapter 6: BGP and Internet Connectivity

Whether or not to use BGP is the first topic discussed in Chapter 6. Different types of ISP connections are covered. The chapter gives an overview of BGP operation and basic configuration. BGP path selection is covered, along with ways to influence the path selection and ways to filter routes. Additionally, ways to verify BGP operation are shown.

Chapter 7: Branch Office Connectivity

Chapter 7 considers branch office routing design and implementation. This includes a description of DSL with PPPoA, and IPsec VPNs. Dynamic routing over various types of tunnels is covered, and a GRE tunnel configuration is shown. The chapter ends with a look at EIGRP load sharing.

Chapter 8: Mobile Worker Connectivity

Chapter 8 covers the network changes necessary to allow mobile workers to connect to the corporate network. The services needed to enabled this are discussed, including NAT, DHCP, firewalling, VPNs, and routing.

Chapter 9: IPv6 Introduction

This rather in-depth introduction to IPv6 covers the IPv6 address format, ways for hosts to acquire their addresses, and IPv6 routing for RIP, EIGRP, OSPF, and BGP. It also includes strategies for integrating IPv4 and IPv6 such as various types of tunnels. IPv6 operation over various types of links is additionally covered.

ROUTE

CHAPTER 1

Planning for Complex Networks

Network Design Models

Today's networks typically include voice, video, network management, mission-critical, and routing traffic in addition to bulk user traffic. Each type of traffic has different performance (bandwidth, delay, and jitter) and security requirements. Network design models provide a framework for integrating the many different types of traffic into the network.

Over the years, several models have been used to help describe how a complex network functions. These models are useful for designing a network and for understanding traffic flow within a more complex network. This section covers three models: the traditional Hierarchical Model, the Enterprise Composite Model, and the Cisco Enterprise Model.

Hierarchical Design Model

Network designers used the three-level *Hierarchical Design Model* for years. This older model provided a high-level idea of how a reliable network might be conceived, but it was largely conceptual because it didn't provide specific guidance. Figure 1-1 shows the Hierarchical Design Model.

This is a simple drawing of how the three-layer model might be built out for a campus network. A distribution Layer-3 switch is used for each building on campus, tying together the access switches on the floors. The core switches link the various buildings together.

This same three-layer hierarchy can be used in the WAN with a central headquarters, division headquarters, and units.

Figure 1-1 Hierarchical Design Model

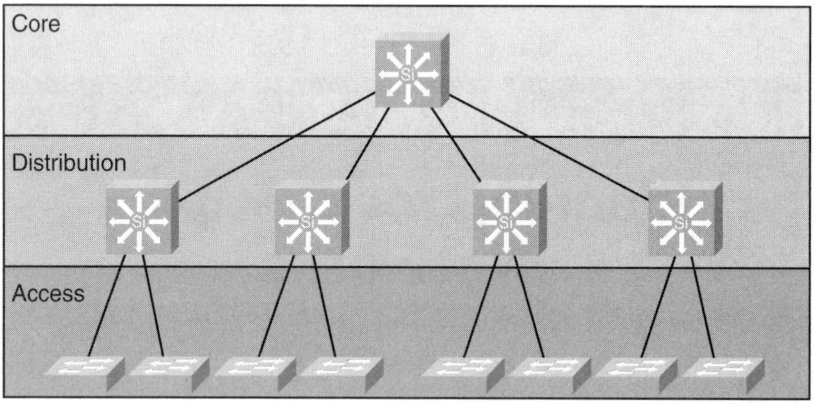

The layers break a network in the following way:

- **Access layer:** Provides network access to workgroup end stations.
- **Distribution layer:** Intermediate devices provide connectivity based on policies.
- **Core layer:** Provides a high-speed switched path between distribution elements.

Redundant distribution and core devices, with connections, make the model more fault-tolerant. This early model was a good starting point, but it failed to address key issues, such as

- Where do wireless devices fit in?
- How should Internet access and security be provisioned?
- How do you account for remote access, such as dial-up or VPN?
- Where should workgroup and enterprise services be located?

Enterprise Composite Model

A newer Cisco model—the Enterprise Composite Model—is significantly more complex and attempts to address the shortcomings of the Hierarchical

Design Model by expanding the older version and making specific recommendations about how and where certain network functions should be implemented. This model is a component of the Cisco Security Architecture for Enterprise (SAFE) Reference Architecture.

The Enterprise Model is broken into three large sections:

- **Enterprise Campus:** Switches that make up a LAN
- **Enterprise Edge:** The portion of the enterprise network connected to the larger world
- **Service Provider Edge:** The different public networks that are attached

The Enterprise Campus, as shown in Figure 1-2, looks like the old Hierarchical Design Model with added details. It features six sections:

- **Campus Backbone:** The core of the LAN
- **Building Distribution:** Connects subnets/VLANs and applies policy
- **Building Access:** Connects users to network
- **Management:** An out-of-band network to access and manage the devices
- **Edge Distribution:** A distribution layer out to the WAN
- **Server Farm:** For Enterprise services

The Enterprise Edge, as shown in Figure 1-3, details the connections from the campus to the WAN and includes

- E-commerce
- Internet connectivity
- Remote access
- WAN

Figure 1-2 Enterprise Campus

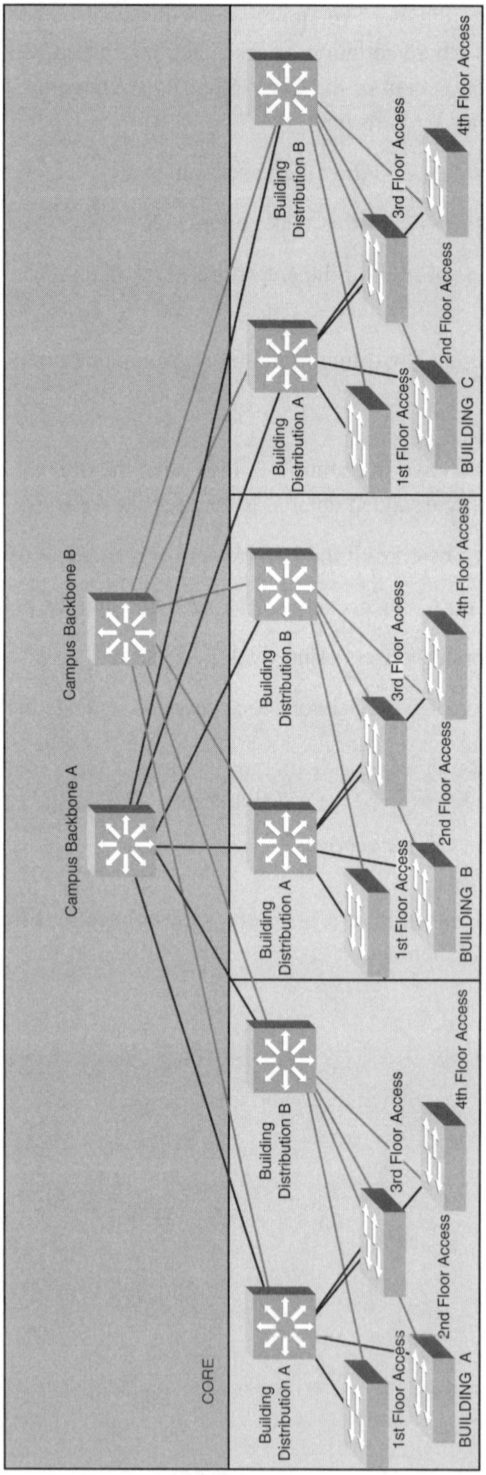

Figure 1-3 Enterprise Edge

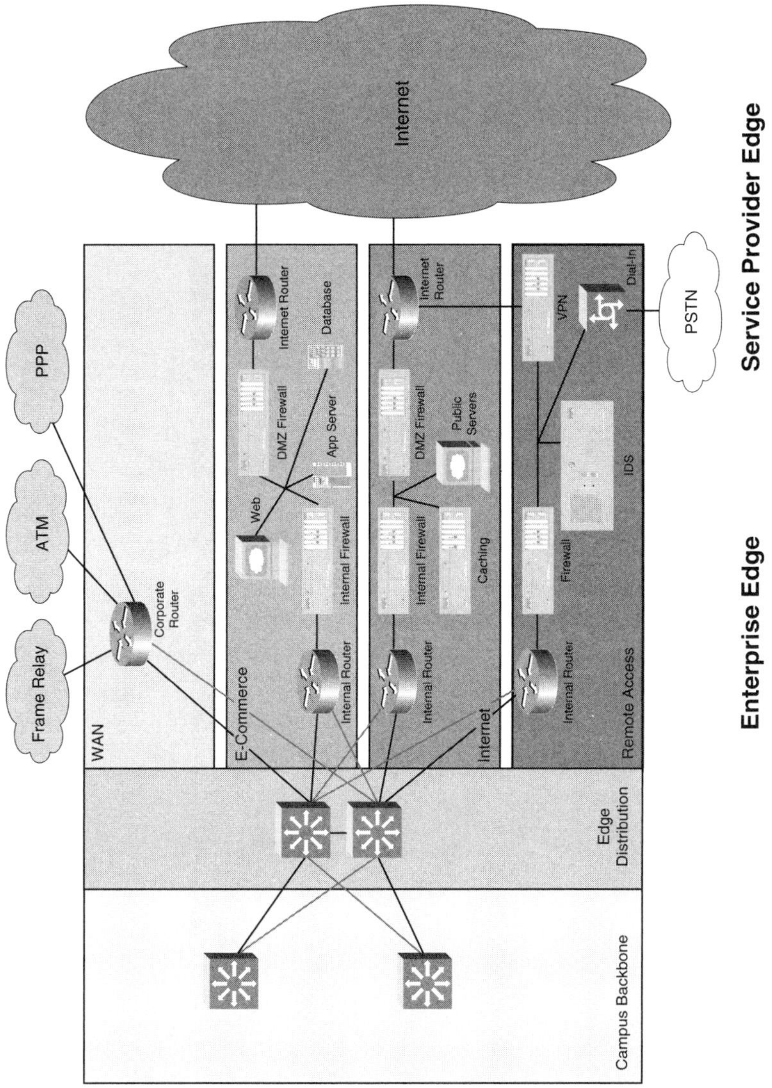

The Service Provider Edge is just a list of the public networks that facilitate wide-area connectivity and include

- Internet service provider (ISP)
- Public switched telephone network (PSTN)
- Frame Relay, ATM, and PPP

Figure 1-4 puts together the various pieces: Campus, Enterprise Edge, and Service Provider Edge. Security implemented on this model is described in the Cisco SAFE blueprint.

Figure 1-4 Enterprise Composite Model

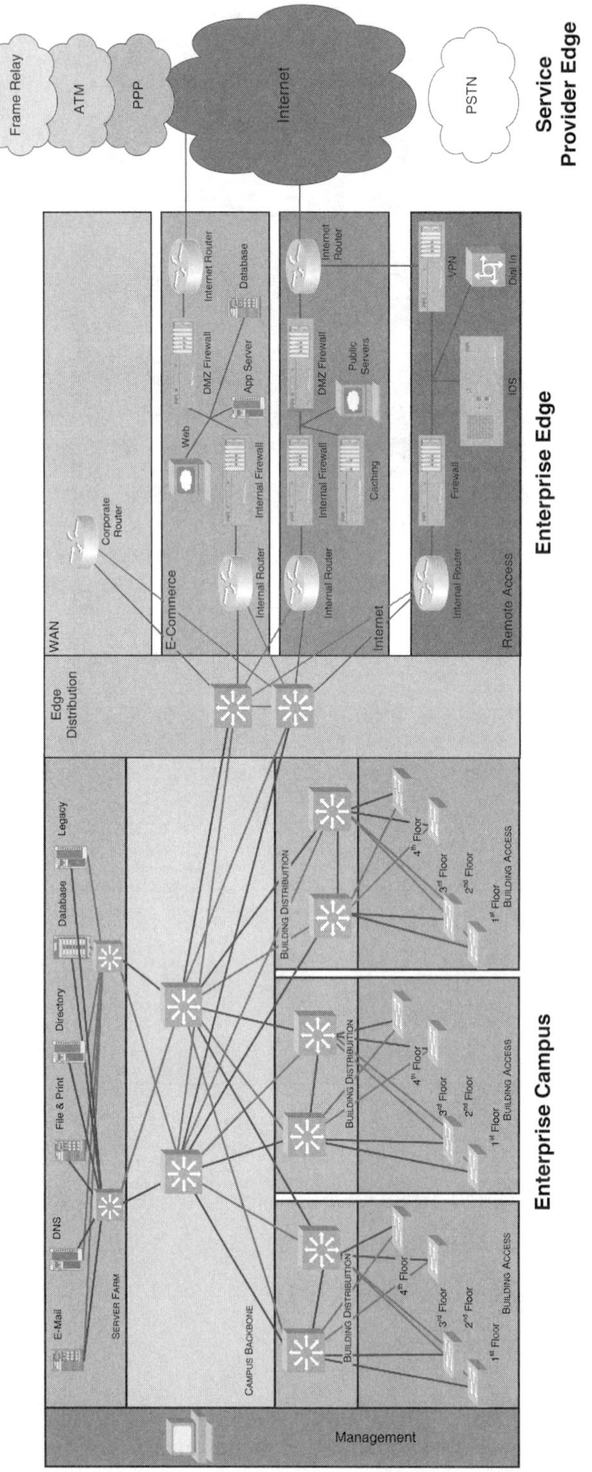

Cisco Enterprise Architecture

The Cisco Enterprise Architecture attempts to describe how all the network components integrate and work together. It includes Campus, Data Center, Branch, WAN, and Teleworker components.

The Campus Architecture component is basically the same as in the Composite model. It includes routing and switching integrated with technologies such as IP telephony and is designed for high availability with redundant links and devices. It integrates security features and provides QoS to ensure application performance. It is flexible enough to add advanced technologies such as VPNs, tunnels, and authentication management.

The Data Center component provides a centralized, scalable architecture that enables virtualization, server and application access, load balancing, and user services. Redundant data centers might be used to provide backup and business continuity.

The Branch Architecture extends enterprise services to remote offices. Network monitoring and management is centralized. Branch networks include access to enterprise-level services such as converged voice and video, security, and application WAN optimization. Resiliency is obtained through backup local call processing, VPNs, redundant WAN links, and application content caching.

The WAN component provides data, voice, and video content to enterprise users any time and any place. QoS, SLAs, and encryption ensure a high-quality secure delivery of resources. It uses IPsec or MPLS VPNs over Layer 2 or Layer 3 WANs, with either a hub-and-spoke or mesh topology.

Teleworker Architecture describes how voice and data are delivered securely to remote small or home office users. It leverages a standard broadband connection, combined with VPN and identity-based access. An IP phone can also be used.

SONA and IIN

Modern converged networks include different traffic types, each with unique requirements for security, QoS, transmission capacity, and delay. These include

- Voice signaling and bearer
- Core application traffic, such as Enterprise Resource Planning (ERP) or Customer Relationship Management (CRM)

- Database transactions
- Multicast multimedia
- Network management
- Other traffic, such as web pages, email, and file transfer

Cisco routers can implement filtering, compression, prioritization, and policing. Except for filtering, these capabilities are referred to collectively as QoS.

Although QoS is a powerful tool, it is not the only way to address bandwidth shortage. Cisco espouses an idea called the Intelligent Information Network (IIN).

IIN describes an evolutionary vision of a network that integrates network and application functionality cooperatively and enables the network to be smart about how it handles traffic to minimize the footprint of applications. IIN is built on top of the Enterprise Composite Model and describes structures overlaid on to the Composite design as needed in three phases.

Phase 1, "Integrated Transport," describes a converged network, which is built along the lines of the Composite model and based on open standards. This is the phase that the industry has been transitioning to recently. The Cisco Integrated Services Routers (ISR) are an example of this trend.

Phase 2, "Integrated Services," attempts to virtualize resources, such as servers, storage, and network access. It is a move to an "on-demand" model.

By "virtualize," Cisco means that the services are not associated with a particular device or location. Instead, many services can reside in one device to ease management, or many devices can provide one service. An ISR brings together routing, switching, voice, security, and wireless. It is an example of many services existing on one device. A load balancer, which makes many servers look like one, is an example of one service residing on many devices.

VRFs are an example of taking one resource and making it look like many. Some versions of IOS are capable of having a router present itself as many virtual router (VRF) instances, allowing your company to deliver different logical topologies on the same physical infrastructure. Server virtualization is another example. The classic example of taking one resource and making it appear to be many resources is the use of a virtual LAN (VLAN) and a virtual storage area network (VSAN).

Virtualization provides flexibility in configuration and management.

Phase 3, "Integrated Applications," uses application-oriented networking (AON) to make the network application-aware and to enables the network to actively participate in service delivery.

An example of this Phase 3 IIN systems approach to service delivery is Network Admission Control (NAC). Before NAC, authentication, VLAN assignment, and antivirus updates were separately managed. With NAC in place, the network can check the policy stance of a client and admit, deny, or remediate based on policies.

IIN enables the network to deconstruct packets, parse fields, and take actions based on the values it finds. An ISR equipped with an AON blade might be set up to route traffic from a business partner. The AON blade handles many functions, including examining traffic, recognizing an application, and rebuilding XML files in memory. Corrupted XML fields might represent an attack (called *schema poisoning*), and the AON blade can react by blocking that source from further communication. In this example, routing, an awareness of the application data flow, and security are all combined to enable the network to contribute to the success of the application.

Services-Oriented Network Architecture (SONA) applies the IIN ideal to Enterprise networks. SONA breaks down the IIN functions into three layers:

- **Network Infrastructure:** Hierarchical converged network and attached end systems
- **Interactive Services:** Resources allocated to applications
- **Applications:** Includes business policy and logic

Understanding Routing Protocols

Routing protocols pass information about the structure of the network between routers. Cisco routers support multiple routing protocols, but the ROUTE exam covers only EIGRP, OSPF, and BGP. This section compares routing protocols and calls out some key differences between them.

Administrative Distance

Cisco routers are capable of supporting several IP routing protocols concurrently. When identical prefixes are learned from two or more separate sources, Administrative Distance (AD) is used to discriminate between the paths. AD is a poor choice of words; *risk-factor* is a more descriptive name. All other things being equal, routers choose paths advertised by the protocol with the lowest AD. AD can be manually adjusted.

Table 1-1 lists the default values for various routing protocols.

Table 1-1 Routing Protocols and Their Default Administrative Distance

Information Source	AD
Connected	0
Static	1
External BGP (Border Gateway Protocol)	20
Internal EIGRP (Enhanced IGRP)	90
IGRP (Internet Gateway Routing Protocol)	100
OSPF (Open Shortest Path First)	110
IS-IS (Intermediate System to Intermediate System)	115
RIP (Routing Information Protocol)	120
ODR (On Demand Routing)	160
External EIGRP	170
Internal BGP	200
Unknown	255

Routing Protocol Characteristics

Two things should always be considered in choosing a routing protocol: fast convergence speed and support for VLSM. EIGRP, OSPF, and BGP all meet these criteria. There are important distinctions between them, as described here:

- EIGRP is proprietary, so it can be used only in an all-Cisco network; however, it is simple for network staff to configure and support.
- OSPF is an open standard, but it is a bit more difficult for network staff to implement and support.
- BGP is also an open standard but is typically used to exchange routes with routers external to your network. It can be very complex to implement, and fewer network engineers understand it well.

Table 1-2 compares routing protocols.

Table 1-2 Comparison of Routing Protocols

Property	EIGRP	OSPF	BGP
Method	Advanced distance vector	Link state	Path vector
Summarization	Auto and manual	Manual	Auto and Manual
VLSM	Yes	Yes	Yes
Convergence Speed	Very fast	Fast	Slow
Timers: Update (hello/dead)	Triggered (LAN 5/15, WAN 60/180)	Triggered, but LSA refreshes every 30 minutes (NBMA 30/120, LAN 10/40)	Triggered (60/180)
Network Size	Large	Large	Very large

Building the Routing Table

The router builds a routing table by ruling out invalid routes and considering the remaining advertisements. The procedure is

1. For each route received, verify the next hop. If invalid, discard the route.
2. If multiple identical, valid routes are received by a routing protocol, choose the lowest metric.
3. Routes are identical if they advertise the same prefix and mask, so 192.168.0.0/16 and 192.168.0.0/24 are separate paths and are each placed into the routing table.
4. If more than one specific valid route is advertised by different routing protocols, choose the path with the lowest AD.

Choosing a Route

Routers look at the routing table to decide how to forward a packet. They look for a match to the destination IP address. Rarely will a route match the destination IP address exactly, so the router looks for the longest match. For instance, suppose a packet is bound for the IP address 10.1.1.1. The routing table has a route for 10.1.0.0/16, one for 10.1.1.0/24, and a default route of 0.0.0.0. The default route matches 0 bits of the destination address, the 10.1.0.0 route matches 16 bits of the destination address, and the 10.1.1.0

route matches 24 bits of the destination address. The 10.1.1.0 route is the longest match, so it will be used to forward the packet.

Planning a Routing Implementation

It is critical to take a structured approach to planning a routing implementation and to document thoroughly once you are done. Taking an ad-hoc approach could lead to network instability, suboptimal routing, or scalability problems.

Four commonly used models include

- **Cisco Lifestyle Services:** Uses the PPDIOO model (Prepare, Plan, Design, Implement, Operate, and Optimize.) Network engineers at the CCNP level are involved with the implementation planning during the Design phase, and the Implementation itself during the Implement phase.
- **IT Infrastructure Library (ITIL):** Emphasizes business requirements and processes as they relate to IT. Implementation and implementation planning are part of its best practices.
- **Fault, Configuration, Accounting, Performance, and Security (FCAPS):** Has five network management categories. Implementation and implementation planning are under the Configuration management category.
- **Telecommunications Management Network (TMN):** Based on the FCAPS model. Implementation and implementation planning are one of its building blocks.

Each approach includes identifying requirements, creating an implementation plan, implementing the changes, verifying your work, and then documenting it.

Creating an Implementation Plan

To create an implementation plan you need to know what the network looks like now, and what it should look like when you are done. This involves gathering information about the current network parameters such as IP addressing, physical connectivity, routing configuration, and equipment. Compare the current state to what is required. Be sure to include any site-specific requirements and any dependencies on the existing network.

An implementation plan includes most of the following, some of which might be site-specific:

- A checklist of tasks to be done
- Tools and resources needed
- The schedule of work, coordinated with all needed resources
- Device configurations
- Verification processes and tests

Creating Implementation Documentation

Documentation should be kept up-to-date, accurate, and accessible. It includes network information, tools and resources used, implementation tasks, verification methods, device configurations, performance measurements, and possibly screen shots or pictures.

CHAPTER 2
EIGRP

EIGRP Overview

Enhanced Interior Gateway Routing Protocol (EIGRP) is a Cisco proprietary, advanced distance vector, classless routing protocol that uses a complex metric based on bandwidth and delay. The following are some features of EIGRP:

- Fast convergence.
- Support for VLSM.
- Partial updates conserve network bandwidth.
- Support for IP, AppleTalk, and IPX.
- Runs directly over IP, using protocol number 88.
- Support for all Layer 2 (data link layer) protocols and topologies.
- Sophisticated metric that supports load-balancing across unequal-cost paths.
- Use of multicast (and unicast where appropriate) instead of broadcasts.
- Support for authentication.
- Manual summarization at any interface.
- Uses multicast 224.0.0.10.

EIGRP's function is controlled by four key technologies:

- **Neighbor discovery and maintenance:** Periodic hello messages
- **The Reliable Transport Protocol (RTP):** Controls sending, tracking, and acknowledging EIGRP messages
- **Diffusing Update Algorithm (DUAL):** Determines the best loop-free route
- **Protocol-independent modules (PDM):** Modules are "plug-ins" for IP, IPX, and AppleTalk versions of EIGRP

EIGRP uses three tables:

- The neighbor table is built from EIGRP hellos and used for reliable delivery.
- The topology table contains EIGRP routing information for best paths and loop-free alternatives.
- EIGRP places best routes from its topology table into the common routing table.

EIGRP Messages

EIGRP uses various message types to initiate and maintain neighbor relationships, and to maintain an accurate routing table. It is designed to conserve bandwidth and router resources by sending messages only when needed and only to those neighbors that need to receive them.

Packet Types

EIGRP uses five packet types:

- **Hello:** Identifies neighbors and serves as a keepalive mechanism
- **Update:** Reliably sends route information
- **Query:** Reliably requests specific route information
- **Reply:** Reliably responds to a query
- **ACK:** Acknowledgment

EIGRP is reliable, but hellos and ACKs are not acknowledged. The acknowledgment to a query is a reply.

If a reliable packet is not acknowledged, EIGRP periodically retransmits the packet to the nonresponding neighbor as a unicast. EIGRP has a window size of one, so no other traffic is sent to this neighbor until it responds. After 16 unacknowledged retransmissions, the neighbor is removed from the neighbor table.

Neighbor Discovery and Route Exchange

When EIGRP first starts, it uses hellos to build a neighbor table. Neighbors are directly attached routers that have a matching AS number and k values.

(The timers don't have to agree.) The process of neighbor discovery and route exchange between two EIGRP routers is as follows:

Step 1. Router A sends out a hello.

Step 2. Router B sends back a hello and an update. The update contains routing information.

Step 3. Router A acknowledges the update.

Step 4. Router A sends its update.

Step 5. Router B acknowledges.

When two routers are EIGRP neighbors, they use hellos between them as keepalives. Additional route information is sent only if a route is lost or a new route is discovered. A neighbor is considered lost if no hello is received within three hello periods (called the *hold time*). The default hello/hold timers are as follows:

- 5 seconds/15 seconds for multipoint circuits with bandwidth greater than T1 and for point-to-point media
- 60 seconds/180 seconds for multipoint circuits with bandwidth less than or equal to T1

The exchange process can be viewed using **debug ip eigrp packets**, and the update process can be seen using **debug ip eigrp**. The neighbor table can be seen with the command **show ip eigrp neighbors**.

EIGRP Route Selection

An EIGRP router receives advertisements from each neighbor listing the advertised distance (AD) and feasible distance (FD) to a route. The AD is the metric from the neighbor to the network. FD is the metric from this router, through the neighbor, to the destination network.

EIGRP Metric

The EIGRP metric is shown in Figure 2-1.

Figure 2-1 EIGRP Metric

$$metric = 256(k1 \times \frac{10^7}{BW_{\min}} + \frac{k2 \times BW_{\min}}{256 - load} + k3 \times \sum delays)(\frac{k5}{reliability + k4})$$

The k values are constants. Their default values are k1 = 1, k2 = 0, k3 = 1, k4 = 0, and k5 = 0. If k5 = 0, the final part of the equation (k5 / [rel + k4]) is ignored.

BW^{min} is the minimum bandwidth along the path—the choke point bandwidth.

Delay values are associated with each interface. The sum of the delays (in tens of microseconds) is used in the equation.

Taking the default k values into account, the equation simplifies to the one shown in Figure 2-2.

Figure 2-2 EIGRP Metric Simplified

$$metric = 256(\frac{10^7}{BW_{\min}} + \sum delays)$$

If default k values are used, this works out to be 256 (BW + cumulative delay).

Bandwidth is the largest contributor to the metric. The delay value enables us to choose a more direct path when bandwidth is equivalent.

Diffusing Update Algorithm (DUAL)

DUAL is the algorithm used by EIGRP to choose best paths by looking at AD and FD. The path with the lowest metric is called the *successor* path. EIGRP paths with a lower AD than the FD of the successor path are guaranteed loop-free and called *feasible successors*. If the successor path is lost, the router can use the feasible successor immediately without risk of loops.

After the router has chosen a path to a network, it is *passive* for that route. If a successor path is lost and no feasible successor is identified, the router sends out queries on all interfaces in an attempt to identify an alternate path. It is *active* for that route. No successor can be chosen until the router receives a reply to all queries. If a reply is missing for 3 minutes, the router becomes *stuck in active (SIA)*. In that case, it resets the neighbor relationship with the neighbor that did not reply.

Three common causes for SIA routes are

- CPU or memory usage is so high on the neighbor that it cannot process the query or reply.

- The link between the routers drops packets. Enough packets get through to maintain the neighbor relationship, but some queries or replies are dropped.
- Unidirectional link, so the router never receives packets from its neighbor.

Route Selection Example

The following diagrams show EIGRP advertisements to R3 and R5 about a destination network connected to R1. In Figure 2-3, R5 chooses R4 as the successor path because it offers the lowest feasible distance. The AD from R3 indicates that passing traffic through R3 will not loop, so R3 is a feasible successor.

Figure 2-3 EIGRP Path Selection, Part One

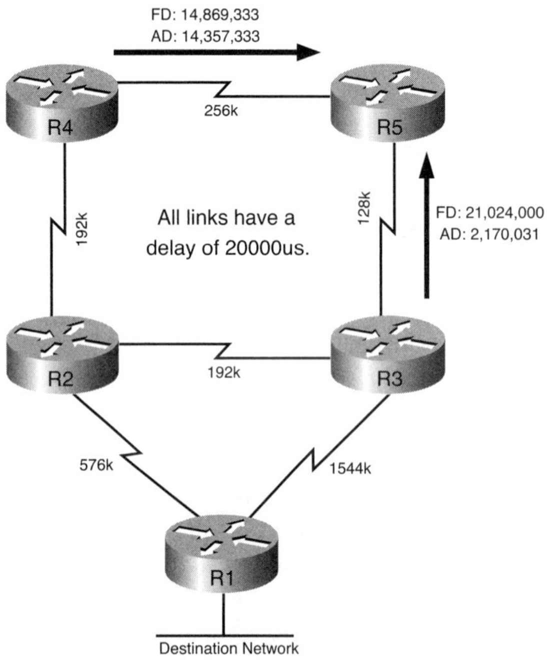

How does R3 choose its path? Figure 2-4 shows the path selection process for R3.

Figure 2-4 EIGRP Path Selection, Part Two

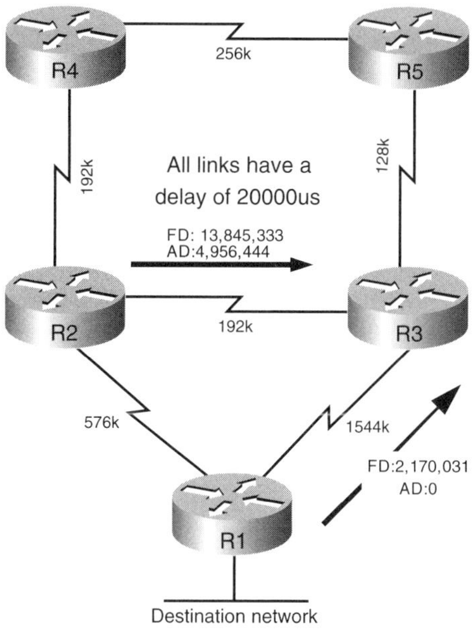

R1 will be its successor because it has the lowest metric. However, no feasible successor exists because R2's AD is greater than the successor path metric. If the direct path to R1 is lost, R3 has to query its neighbors to discover an alternative path. It must wait to hear back from R2 and R5 and will ultimately decide that R2 is the new successor.

Planning an EIGRP Implementation

When planning an EIGRP implementation, gather the following information:

- **Current network setup and future requirements:** Document the IP addressing used and the network topology, including links types, bandwidth, and utilization. A good IP addressing design allows summarization at various points in the network.
- **Network design:** Although EIGRP does not require a hierarchical network design, it can perform more efficiently within that type of network.
- **Plans for EIGRP scaling options:** These would include summarization, stub areas, and changes in interface metrics to improve bandwidth utilization.

Your final implementation plan needs to include detailed parameters such as the exact topology, IP networks to be advertised, EIGRP AS number, lists of routers to run EIGRP, and any nondefault metrics to be used. It needs to list implementation tasks for each router in the network. Finally it needs to provide verification tasks for each router such as verifying neighbors, IP routing tables, EIGRP topology tables, and network connectivity.

Basic EIGRP Configuration

EIGRP is configured by entering router configuration mode and identifying the networks within which it should run. When setting up EIGRP, an autonomous system number must be used (7 is used in the example). Autonomous system numbers must agree for two routers to form a neighbor relationship and to exchange routes.

```
Router(config)# router eigrp 7
Router(config-router)# network 192.168.1.0
```

The wildcard mask option can be used with the network command to more precisely identify EIGRP interfaces. For instance, if a router has two interfaces—fa0/0 (192.168.1.1/27) and fa0/1 (192.168.1.33/27)—and needs to run EIGRP only on fa0/0, the following command can be used:

```
Router(config-router)# network 192.168.1.0 0.0.0.1
```

In this command, a wildcard mask of 0.0.0.1 matches only two IP addresses in network 192.168.1.0–192.168.1.0 and 192.168.1.1. Therefore, only interface fa0/0 is included in EIGRP routing.

To ensure that the correct metric is calculated, or to influence the metric, you might want to configure the bandwidth on the interface. Use the interface command:

```
R1(config)# interface s0/0/0
R1(config-if)# bandwidth kbps
```

Creating an EIGRP Default Route

Figure 2-5 shows a simple two-router network. You can configure EIGRP on R1 to advertise a default route to R3 in these ways:

- R1 can specify a default network:

  ```
  R1(config)# ip default-network 10.0.0.0
  ```

  R3 now sees a default network with a next hop of R1.

- Create a static default route and then include network 0.0.0.0 in EIGRP:

```
R1(config)# ip route 0.0.0.0 0.0.0.0 10.1.1.2
R1(config)# router eigrp 7
R1(config-router)# network 0.0.0.0
```

Figure 2-5 EIGRP Default Route

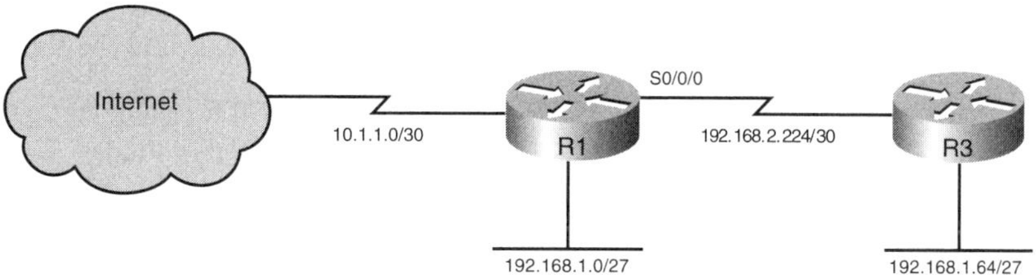

Verify and Troubleshoot EIGRP

The most straightforward way to troubleshoot EIGRP is to inspect the routing table, **show ip route**. To filter the routing table and show only the routes learned from EIGRP, use the **show ip route eigrp** command. The **show ip protocols** command verifies autonomous system, timer values, identified networks, and EIGRP neighbors (routing information sources).

The command **show ip eigrp topology** shows the EIGRP topology table and identifies successors and feasible successors. Use **show ip eigrp neighbors** to verify that the correct routers are neighbors, and use **show ip eigrp traffic** to show the amount and types of EIGRP messages. The command **show ip eigrp interfaces** lists the interfaces participating in EIGRP and any neighbors found out those interfaces, along with some other statistics.

EIGRP Across a WAN

EIGRP can be used across many types of WAN links. This section examines how it operates over some of them.

EIGRP over EoMPLS

MPLS can provide either a Layer 2 or a Layer 3 connection. In MPLS terminology, your WAN edge routers are called CE (customer edge) routers, and the ISP's WAN edge routers are called PE (provider edge) routers. Within the ISP's network are P (provider) routers, but they should not be visible to the CE.

Ethernet over MPLS (EoMPLS) leverages Any Transport over MPLS (AToM) to provide a Layer 2 connection such as Metro Ethernet. With EoMPLS, the CE routers appear to have a point-to-point Ethernet connection across the WAN. In reality, each CE router has an Ethernet connection to its local PE router.

Figure 2-6 shows how this works. The PE1 router receives Ethernet frames from CE1, encapsulates them into an MPLS packet, and then forwards them across the WAN to PE2, which is the local router connected to CE2. PE2 decapsulates the packet, rebuilds the Ethernet frame, and sends it to the CE2.

It is important to understand that CE1 and CE2 build an EIGRP neighbor relationship with each other. The ISP routers are not involved in routing with the CE routers. Additionally, the PE routers do not learn any MAC addresses or participate in Spanning Tree.

Figure 2-6 Using EIGRP with EoMPLS

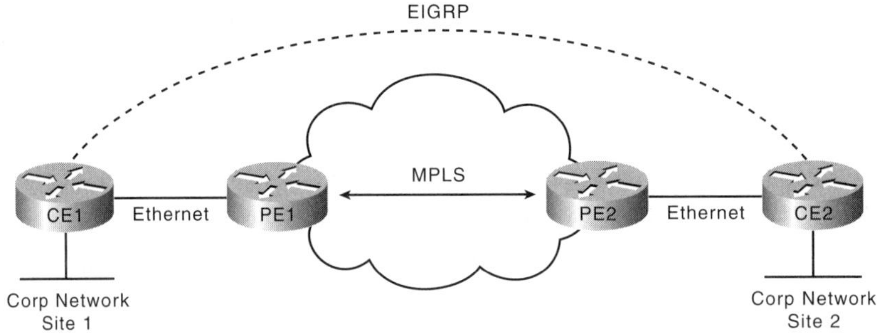

EIGRP over MPLS

PE routers are involved in routing when you use EIGRP over Layer 3 MPLS VPNs, however. The connection between the CE and PE routers is a Layer 3 connection. Each connected PE and CE router are EIGRP neighbors. The PE router is just another neighbor to the CE router; it is not aware of the MPLS network or the ISP's P routers.

In Figure 2-7, CE1 creates an EIGRP neighbor relationship with PE1. CE1 sends routing updates about its networks to PE1, which installs the routes in the correct Virtual Routing and Forwarding (VRF) table and then transmits them across the WAN as MPLS packets to PE2. PE2 is an EIGRP neighbor to CE2, so it forwards the route advertisements as normal EIGRP updates.

When using EIGRP over MPLS, the customer and the provider need to use the same basic EIGRP configuration such as AS number and authentication.

Figure 2-7 EIGRP with MPLS

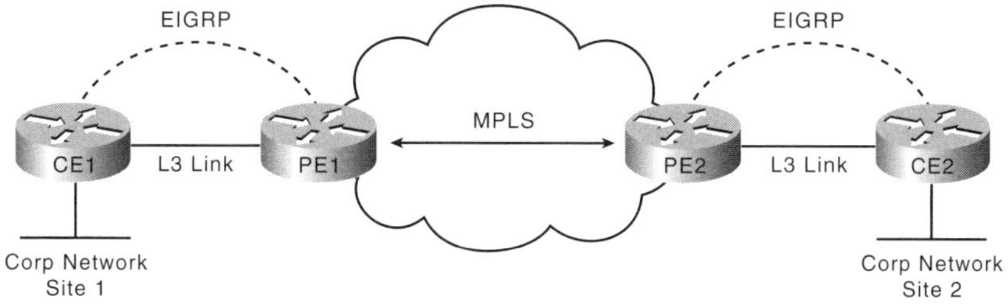

EIGRP over Frame Relay

One issue with using EIGRP over Frame Relay is that one physical interface can support multiple logical connections, each identified by a Data Link Connection Identifier (DLCI). These are Layer 2 connections and must be mapped to a Layer 3 neighbor IP address. This mapping can be done either dynamically or statically. Multipoint interfaces are used in partial and full mesh topologies.

Dynamic mapping uses Inverse ARP. Routers form EIGRP neighbor adjacencies only with routers that they connect to via a Frame Relay virtual circuit (VC). Static mapping requires manual configuration under each interface but enables routers without VC connections to become neighbors. The static mapping command is given under interface configuration mode:

```
frame-relay map ip remote-ip-address local-dlci broadcast
```

The **broadcast** keyword is required because Frame Relay is, by default, a nonbroadcast medium. Static mapping can be used with both physical multipoint interfaces and subinterfaces. Note that a multipoint interface stays up if one DLCI is active, so a neighbor loss might not be detected until the hold timer expires.

Frame Relay can emulate physical point-to-point links by using point-to-point subinterfaces. This is used in a hub-and-spoke topology. Neighbor loss is detected much more quickly on point-to-point links for two reasons:

- The default timers are shorter, 5 second hold timer and 15 second dead timer.
- The subinterface goes down when its associated DLCI goes down.

WAN Bandwidth

By default, EIGRP limits itself to bursting to half the link bandwidth. This limit is configurable per interface using the **ip bandwidth-percent** command. The following example assumes EIGRP AS 7 and limits EIGRP to one quarter of the link bandwidth:

```
Router(config)# int s0/0/0
Router(config-if)# ip bandwidth-percent eigrp 7 25
```

The real issue with WAN links is that the router assumes that each link has 1544 kbps bandwidth. If interface Serial0/0/0 is attached to a 128 k fractional T1, EIGRP assumes it can burst to 768 k and could overwhelm the line. This is rectified by correctly identifying link bandwidth:

```
Router (config)# int serial 0/0/0
Router (config-if)# bandwidth 128
```

Figure 2-8 shows a situation in which these techniques can be combined: Frame Relay.

Figure 2-8 EIGRP with Frame Relay

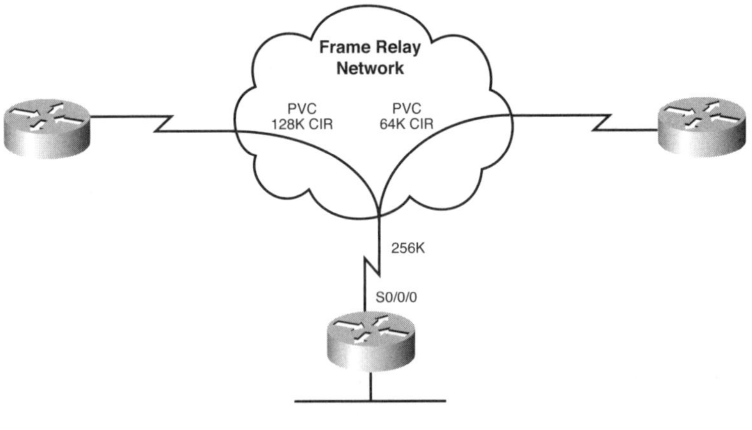

In this example, R1 has a 256 kbps connection to the Frame Relay network and two permanent virtual circuits (PVCs) with committed information rates (CIR) of 128 Kpbs and 64 Kbps. EIGRP divides the interface bandwidth evenly between the number of neighbors on that interface. What value should be used for the interface bandwidth in this case? The usual suggestion is to use the CIR, but the two PVCs have different CIRs. You can use the bandwidth-percent command to allow SNMP reporting of the true bandwidth value, while adjusting the interface burst rate to 25 percent, or 64 kbps.

```
R1(config)# int serial 0/0/0
R1 (config-if)# bandwidth 256
R1 (config-if)# ip bandwidth-percent eigrp 7 25
```

A better solution is to use point-to-point subinterfaces and identify bandwidth separately. In the following example, s0/0/0.1 bursts to 64 k, and s0/0/0.2 bursts to 32 k, using EIGRP's default value of half the bandwidth.

```
R1(config)# int serial 0/0/0.1 point-to-point
R1(config-if)# bandwidth 128
R1(config-if)# frame-relay interface-dlci 100
!
R1(config)# int serial 0/0/0.2 point-to-point
R1(config-if)# bandwidth 64
R1(config-if)# frame-relay interface-dlci 101
```

In cases where the hub interface bandwidth is oversubscribed, it might be necessary to set bandwidth for each subinterface arbitrarily low and then specify an EIGRP bandwidth percent value over 100 to allow EIGRP to use half the PVC bandwidth.

Customizing the EIGRP Configuration

EIGRP provides some ways to customize its operation, such as passive interface, unicast neighbors, route summarization, unequal-metric load balancing, and authentication. This section describes how to configure these.

Passive Interface

The **passive-interface** command prevents either routing updates or hello messages from being sent out an interface. RIP does not send updates when it enabled; EIGRP and OSPF do not send hellos, and thus they don't discover neighbors or form an adjacency out that interface. To disable the protocol on one interface, use the routing protocol configuration command **passive-interface** *interface*. To turn off the protocol on all interfaces, use **passive-interface default**. You can then use **no passive-interface** *interface* for the ones that should run the protocol, as shown here:

```
Router(config)# router eigrp 7
Router(config-router)# passive-interface default
Router(config-router)# no passive-interface s0/0/0
```

Unicast Neighbors

EIGRP usually uses a multicast to IP address 224.0.0.10 for its messages. You can configure it to use a unicast address instead with the routing protocol configuration command **neighbor** *ip-address*. The IP address must be in the same subnet as one of the router's own interfaces.

Summarization

EIGRP defaults to automatically summarizing at classful network boundaries. Automatic summarization is usually disabled using the following command:

```
Router(config-router)# no auto-summary
```

Summaries can be produced manually on any interface. When a summary is produced, a matching route to null0 also becomes active as a loop prevention mechanism. Configure a summary route out a particular interface using the **ip summary-address eigrp** *autonomous_system* command. The following example advertises a default route out FastEthernet0/1 and the summary route 172.16.104.0/22 out Serial0/0/0 for EIGRP AS 7.

```
Router(config)# int fa0/1
Router(config-if)# ip summary-address eigrp 7 0.0.0.0 0.0.0.0
!
Router(config)# int s0/0/0
Router(config-if)# ip summary-address eigrp 7 172.16.104.0
 255.255.252.0
```

Load Balancing

EIGRP, like most IP routing protocols, automatically load balances over equal metric paths. What makes EIGRP unique is that you can configure it to proportionally load balance over *unequal* metric paths. The **variance** command is used to configure load balancing over up to six loop-free paths with a metric lower than the product of the variance and the best metric. Figure 2-9 shows routers advertising a path to the network connected to R1.

By default, R5 uses the path through R4 because it offers the lowest metric (14,869,333). To set up unequal cost load balancing, assign a variance of 2 under the EIGRP process on R5, which multiplies the best metric of 14,869,333 by 2 to get 29,738,666. R5 then uses all loop-free paths with a metric less than 29,738,666, which includes the path through R3. By default, R5 load balances over these paths, sending traffic along each path in proportion to its metric.

```
R5(config)# router eigrp 7
R5(config-router)# variance 2
```

Figure 2-9 EIGRP Unequal-cost Load Balancing

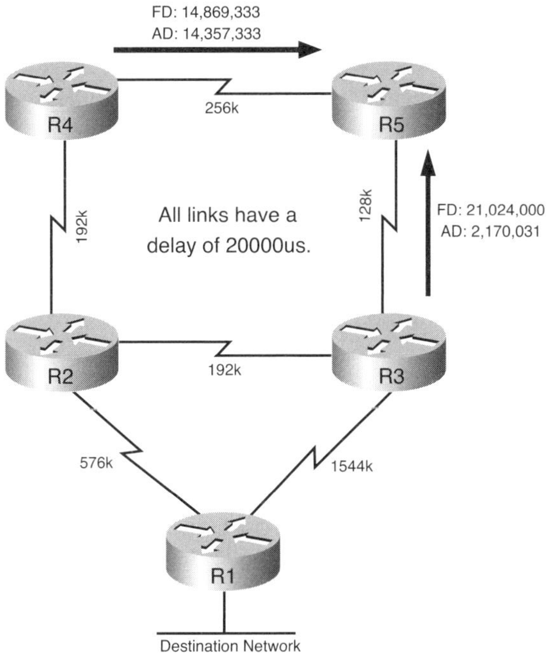

EIGRP Authentication

By default, no authentication is used for any routing protocol. Some protocols, such as RIPv2, IS-IS, and OSPF, can be configured to do simple password authentication between neighboring routers. In this type of authentication, a clear-text password is used. EIGRP does not support simple authentication. However, it can be configured to authenticate each packet exchanged using an MD5 hash created from a preconfigured, shared password. This is more secure than clear text because only the message digest is exchanged, not the password. The password is called the *key*.

EIGRP authenticates each of its packets and verifies the source of each routing update by including the hash in each one. If the hash value does not match, the packet is silently dropped.

To implement EIGRP authentication, first create a plan:

- Look at the current configuration to determine the AS number and interfaces where it will be configured.
- Decide the authentication type. (For EIGRP this must be MD5.)
- Decide the key strings, and how many keys will be used.
- Optionally decide the key lifetimes.

To configure the router for EIGRP authentication, follow these steps:

Step 1. Configure a key chain to group the keys.

Step 2. Configure one or more keys within that key chain. The router checks all inbound packets against the list of keys and uses the first valid one it finds.

Step 3. Configure the password or authentication string for that key. Repeat Steps 2 and 3 to add more keys if desired.

Step 4. Optionally configure a lifetime for the keys within that key chain. If you do this, be sure that the time is synchronized between the two routers.

Step 5. Enable authentication and assign a key chain to an interface.

Step 6. Designate MD5 as the type of authentication.

Example 2-1 shows a router configured with EIGRP authentication. It shows configuring a lifetime for packets sent using key 1 that starts at 10:15 and lasts for 300 seconds. It also shows configuring a lifetime for packets received using key 1 that starts at 10:00 and lasts until 10:05. Router clocks must be synchronized when using lifetimes, so use an NTP server.

Example 2-1 Configuring EIGRP Authentication

```
Router(config)# key chain RTR_Auth
Router(config-keychain)# key 1
Router(config-keychain-key)# key-string mykey
Router(config-keychain-key)# send-lifetime 10:15:00 300
Router(config-keychain-key)# accept-lifetime 10:00:00 10:05:00
!
Router(config)# interface s0/0/0
Router(config-if)# ip authentication mode eigrp 10 md5
Router(config-if)# ip authentication key-chain eigrp 10 RTR_Auth
```

Verify your configuration with the **show key chain** command. **show ip eigrp neighbors** is also useful, as no neighbor relationship will be formed if authentication fails. Using the **debug eigrp packets** command should show packets containing authentication information sent and received, and it enables you to troubleshoot configuration issues. The debug output lists an authentication mismatch message if authentication does not succeed.

EIGRP Scalability

Four factors influence EIGRP's scalability:

- The number of routes that must be exchanged
- The number of routers that must know of a topology change
- The number of alternate routes to a network
- The number of hops from one end of the network to the other (topology depth)

To improve scalability, summarize routes when possible, try to have a network depth of no more than seven hops, and limit the scope of EIGRP queries.

EIGRP Stub

Stub routing is one way to limit queries. A stub router is one that is connected to no more than two neighbors and should never be a transit router. This feature is commonly used in a hub-and-spoke topology. When a router is configured as an EIGRP stub, it notifies its neighbors. The neighbors then do not query that router for a lost route. An EIGRP stub router still receives all routes from its neighbors by default.

Under router configuration mode, use the command **eigrp stub [receive-only|connected|static|summary|redistributed]**. Table 2-1 lists each of the command options and their affect.

Table 2-1 eigrp stub Command Options

Command Option	Affect
`receive-only`	Prevents the router from advertising any networks, including its own. Cannot be combined with any other option.
`connected`	Enables the router to advertise connected routes. These must either be included in a network statement or redistributed into EIGRP. Enabled by default.
`static`	Enables the router to advertise static routes. They must be redistributed into EIGRP before they will be advertised.
`summary`	Enables the router to advertise summary routes, both those created manually and automatically. Enabled by default.
`redistributed`	Allows the router to advertise routes redistributed into EIGRP from another protocol or AS.

Active Process Enhancement

The Active Process Enhancement enables routers to use *SIA-Queries* and *SIA-Replies* to prevent the loss of a neighbor unnecessarily during SIA conditions. A router sends its neighbor a SIA-Query after no reply to a normal query. If the neighbor responds with a SIA-Reply, the router does not terminate the neighbor relationship after 3 minutes, because it knows the neighbor is available.

Graceful Shutdown

Graceful shutdown is another feature that speeds network convergence. Whenever the EIGRP process is shut down, the router sends a "goodbye" message to its neighbors. Ironically, the goodbye message is sent in a "hello" packet. The neighbors can then immediately recalculate any paths that used the router as the next hop, rather than waiting for the hold timer to expire.

CHAPTER 3

OSPF

OSPF Overview

OSPF is an open-standard, classless routing protocol that converges quickly and uses cost as a metric. (Cisco IOS automatically associates cost with bandwidth.)

OSPF is a link-state routing protocol and uses Dijkstra's Shortest Path First (SPF) algorithm to determine its best path to each network. The first responsibility of a link-state router is to create a database that reflects the structure of the network. Link state routing protocols learn more information on the structure of the network than other routing protocols and thus can make more informed routing decisions.

OSPF routers exchange Hellos with each neighbor, learning Router ID (RID) and cost. Neighbor information is kept in the adjacency database.

The router then constructs the appropriate Link State Advertisements (LSA), which include information such as the RIDs of, and cost to, each neighbor. Each router in the routing domain shares its LSAs with all other routers. Each router keeps the complete set of LSAs in a table—the Link State Database (LSDB).

Each router runs the SPF algorithm to compute best paths. It then submits these paths for inclusion in the routing table, or forwarding database.

OSPF Network Structure

OSPF routing domains are broken up into areas. An OSPF network must contain an area 0 and might contain other areas. The SPF algorithm runs within an area, and interarea routes are passed between areas. A two-level hierarchy to OSPF areas exists; area 0 is designed as a transit area, and other areas should be attached directly to area 0 and only to area 0. The link-state database must be identical for each router in an area. OSPF areas typically contain a maximum of 50 routers to 100 routers, depending on network volatility. 3-1 shows a network of five routers that has been divided into three areas: area 0, area 1, and area 2.

Figure 3-1 OSPF Areas

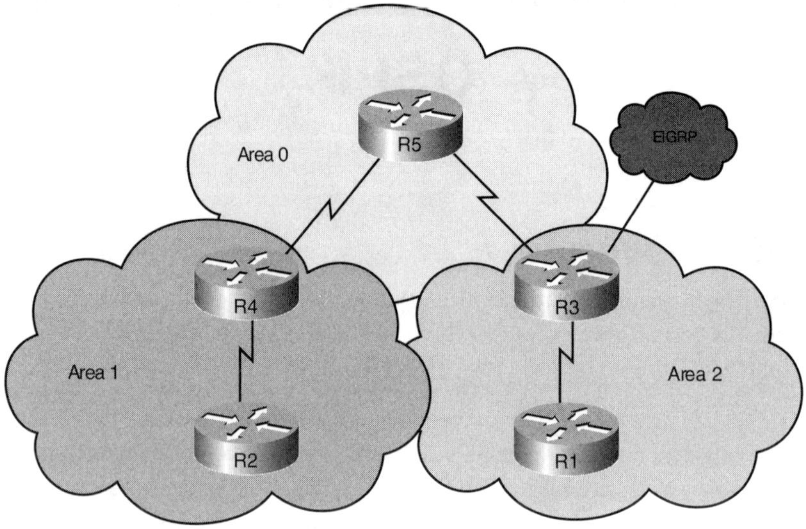

Dividing an OSPF network into areas does the following:

- Minimizes the number of routing table entries
- Contains LSA flooding to a reasonable area
- Minimizes the impact of a topology change
- Enforces the concept of a hierarchical network design

Following are several types of areas:

- **Backbone area:** Area 0, which is attached to every other area.
- **Regular area:** Nonbackbone area; its database contains both internal and external routes.
- **Stub area:** It's database contains only internal routes and a default route.
- **Totally Stubby Area:** Cisco proprietary area designation. Its database contains routes only for its own area and a default route.
- **Not-so-stubby area (NSSA):** Its database contains internal routes, routes redistributed from a connected routing process, and optionally a default route.
- **Totally NSSA:** Cisco proprietary area designation. Its database contains only routes for its own area, routes redistributed from a connected routing process, and a default route.

OSPF defines router roles as well. One router can have multiple roles:

- An internal router has all interfaces in one area. In Figure 3-1, R1, R2, and R5 are all internal area routers. They maintain a link-state database for their own area only.
- Backbone routers have at least one interface assigned to area 0. R3, R4, and R5 are backbone routers.
- An Area Border Router (ABR) has interfaces in two or more areas. In Figure 3-1, R3 and R4 are ABRs. ABRs separates LSA flooding areas, can summarize area routes, and can source default routes. They maintain a link-state database for each area to which they are connected.
- An Autonomous System Boundary Router (ASBR) has interfaces inside and outside the OSPF routing domain. In Figure 3-1, R3 additionally functions as an ASBR because it has an interface in an EIGRP routing domain.

OSPF Metric

By default, Cisco assigns a cost to each interface that is inversely proportional to 100 Mbps (100,000,000 bps). The cost for each link is then accrued as the route advertisement for that link traverses the network. 3-2 shows the default OSPF formula.

Figure 3-2 OSPF Cost Formula

$$\text{Cost} = \frac{100\ \text{Mbps}}{\text{Bandwidth}}$$

The default formula doesn't differentiate between interfaces with speeds faster than 100 Mbps. It assigns the same cost to a Fast Ethernet interface and a Gigabit Ethernet interface, for example. In such cases, the cost formula can be adjusted using the **auto-cost** command under the OSPF routing process. Values for bandwidth (in kbps) up to 4,294,967 are permitted (1 Gbps is shown in the following line):

```
Router(config-router)# auto-cost reference-bandwidth 1000
```

The cost can also be manually assigned under the interface configuration mode. The cost is a 16-bit number, so it can be any value from 1 to 65,535.

```
Router(config-if)# ip ospf cost 27
```

Link State Advertisements (LSA)

Each router maintains a database, called the *link-state database (LSDB)*, containing the latest received LSAs. A separate LSDB is maintained for each area connected to the router.

LSA Operation

Each LSA is numbered with a sequence number, and a timer is run to age out old LSAs. The default timer is 30 minutes.

When a LSA is received, it's compared to the LSDB. If it is new, it is added to the database, and the SPF algorithm is run. If it is from a Router ID that is already in the database, the sequence number is compared, and older LSAs are discarded. If it is a new LSA, it is incorporated in the database, and the SPF algorithm is run. If it is an older LSA, the newer LSA in memory is sent back to whoever sent the old one.

OSPF sequence numbers are 32 bits. The first legal sequence number is 0x80000001. Larger numbers are more recent. The sequence number changes only under two conditions:

- The LSA changes because a route is added or deleted.
- The LSA ages out. (LSA updates are flooded within the area every half hour, even if nothing changes.)

The command **show ip ospf database** shows the age (in seconds) and sequence number for each router.

LSDB Overload Protection

Because each router sends an LSA for each link, routers in large networks might receive—and must process—numerous LSAs. This can tax the router's CPU and memory resources, and adversely affect its other functions. LDSB overload protection monitors the number of LSAs received and placed into the LSDB. If the specified threshold is exceeded for one minute, the router enters the "ignore" state by dropping all adjacencies and clearing the OSPF database. The router resumes OSPF operations after things have been normal for a specified period. Be careful because this feature disrupts routing when invoked.

Configure LSDB overload protection with the OSPF router process command **max-lsa** *maximum-number* [*threshold-percentage*] [**warningonly**][**ignore-time** *minutes*] [**ignore-count** *number*] [**reset-time** *minutes*].

LSA Types

OSPF uses different types of LSAs to advertise different types of routes, such as internal area or external routing domain. Many of these are represented in the routing table with a distinctive prefix. Table 3-1 describes these LSA types.

Table 3-1 OSPF LSA Types

Type	Description	Routing Table Symbol
1	Router LSA. Advertises intra-area routes. Generated by each OSPF router. Flooded only within the area.	O
2	Network LSA. Advertises routers on a multiaccess link. Generated by a DR. Flooded only within the area.	O
3	Summary LSA. Advertises interarea routes. Generated by an ABR. Flooded to adjacent areas.	O IA
4	Summary LSA. Advertises the route to an ASBR. Generated by an ABR. Flooded to adjacent areas.	O IA
5	External LSA. Advertises routes in another routing domain. Generated by an ASBR. Flooded to adjacent areas. E1–metric increases at each router as it is passed through the network. O E2–metric does not increase (this is the default).	O
6	Multicast LSA. Used in multicast OSPF operations.	
7	Not-so-stubby area (NSSA) LSA. Advertises routes in another routing domain. Generated by an ASBR within a not-so-stubby area. N1–metric increases as it is passed through the network. O N2–metric does not increase (default).	O
8	External attributes LSA. Used in OSPF and BGP interworking.	
9,10,11	Opaque LSAs. Used for specific applications, such as OSPF and MPLS interworking.	

OSPF Operation

OSPF uses several different message types to establish and maintain its neighbor relationships and to maintain correct routing information. When preparing for the exam, be sure you understand each OSPF packet type and the OSPF neighbor establishment procedure.

OSPF Packets

OSPF uses five packet types. It docs not use UDP or TCP for transmitting its packets. Instead, it runs directly over IP (IP protocol 89) using an OSPF header. One field in this header identifies the type of packet being carried. The five OSPF packet types follow:

- **Hello:** Identifies neighbors and serves as a keepalive.
- **Link State Request (LSR):** Request for a Link State Update (LSU). Contains the type of LSU requested and the ID of the router requesting it.
- **Database Description (DBD):** A summary of the LSDB, including the RID and sequence number of each LSA in the LSDB.
- **Link State Update (LSU):** Contains a full LSA entry. An LSA includes topology information; for example, the RID of this router and the RID and cost to each neighbor. One LSU can contain multiple LSAs.
- **Link State Acknowledgment (LSAck):** Acknowledges all other OSPF packets (except Hellos).

OSPF traffic is multicast to either of two addresses: 224.0.0.5 for all OSPF routers or 224.0.0.6 for all OSPF DRs.

OSPF Neighbor Relationships

OSPF routers send out periodic multicast packets to introduce themselves to other routers on a link. They become neighbors when they see their own router ID included in the Neighbor field of the Hello from another router. Seeing this tells each router that they have bidirectional communication. In addition, two routers must be on a common subnet for a neighbor relationship to be formed. (Virtual links are sometimes an exception to this rule.)

Certain parameters within the OSPF Hellos must also match for two routers to become neighbors. They include

- Hello/dead timers
- Area ID
- Authentication type and password
- Stub area flag

OSPF routers can be neighbors without being adjacent. Only adjacent neighbors exchange routing updates and synchronize their databases. On a point-to-point link, an adjacency is established between the two routers when they can communicate. On a multiaccess link, each router establishes an adjacency only with the DR and the backup DR (BDR).

Hellos also serve as keepalives. A neighbor is considered lost if no Hello is received within four Hello periods (called the dead time). The default Hello/dead timers are as follows:

- 10 seconds/40 seconds for LAN and point-to-point interfaces
- 30 seconds/120 seconds for nonbroadcast multiaccess (NBMA) interfaces

Establishing Neighbors and Exchanging Routes

The process of neighbor establishment and route exchange between two OSPF routers is as follows:

Step 1. **Down state:** OSPF process not yet started, so no Hellos sent.

Step 2. **Init state:** Router sends Hello packets out all OSPF interfaces.

Step 3. **Two-way state:** Router receives a Hello from another router that contains its own router ID in the neighbor list. All other required elements match, so routers can become neighbors.

Step 4. **Exstart state:** If routers become adjacent (exchange routes), they determine which one starts the exchange process.

Step 5. **Exchange state:** Routers exchange DBDs listing the LSAs in their LSD by RID and sequence number.

Step 6. **Loading state:** Each router compares the DBD received to the contents of its LS database. It then sends a LSR for missing or outdated LSAs. Each router responds to its neighbor's LSR with a Link State Update. Each LSU is acknowledged.

Step 7. **Full state:** The LSDB has been synchronized with the adjacent neighbor.

Planning for OSPF

Planning an OSPF implementation is more stringent than planning for EIGRP, because OSPF has specific network design requirements. Gather the following information:

- **Current network setup and future requirements:** Document the IP addressing used and the network topology, including links types, bandwidth, and utilization. Document the current router utilization. Create an IP addressing design that allows summarization at the ABRs.

- **Network design:** OSPF requires a hierarchical network design. You must create a backbone area (area 0) and normal areas. Area 0 must be contiguous. Normal areas must be connected to area 0 either directly or via a virtual link. Decide where the area boundaries should fall. Ensure that the normal areas have sufficient connectivity to area 0, and that all ABRs have the resources to handle the OSPF traffic in addition to the user traffic.
- **Plans for OSPF scaling options:** These would include summarization and stub areas.

Your final implementation plan needs to include detailed parameters such as the exact topology, IP networks to be advertised, OSPF process number, lists of routers to run OSPF, and any changes needed to the default interface metric. It needs to list implementation tasks for each router in the network. Finally it needs to provide verification tasks for each router such as verifying neighbors, IP routing tables, OSPF topology tables, and network connectivity. Document the new network configurations.

Basic OSPF Configuration

OSPF is configured by entering router configuration mode and identifying the range of interface addresses on which it should run and the areas they are in. When setting up OSPF, a process ID must be used (8 is used in the example), but the process ID does not need to agree on different OSPF devices for them to exchange information. The network statement uses a wildcard mask and can specify any range from a single address to all addresses. Unlike EIGRP, the wildcard mask is not optional. The following example shows a router configured as an ABR. Interfaces falling with the 192.168.1.0 network are placed in area 0, and interfaces falling within the 172.16.1.0 network are placed in area 1.

```
Router(config)# router ospf 8
Router(config-router)# network 192.168.1.0 0.0.0.255 area 0
Router(config-router)# network 172.16.1.0 0.0.0.255 area 1
```

Alternatively, you can enable OSPF directly on an interface, rather than using a network statement. This is especially helpful on unnumbered interfaces and enables more granular control over which interfaces run OSPF.

```
Router(config)# int s 0/0/0
Router(config-if)# ip ospf 8 area 0
```

The **ip ospf area** interface command takes precedence over a **network** command.

Router ID

The SPF algorithm maps the shortest path between a series of nodes. This causes an issue with IP because an IP router is not identified by a single IP address; its interfaces are. For this reason, a single IP address is designated as the "name" of the router: the Router ID (RID).

By default, the RID is the highest loopback IP address. If no loopback addresses are configured, the RID is the highest IP address on an active interface when the OSPF process is started. The RID is selected when OSPF starts and—for reasons of stability—is not changed until OSPF restarts. The OSPF process can be restarted by rebooting or by using the command **clear ip ospf process**. Either choice affects routing in your network for a period of time and should be used only with caution.

A loopback interface is a virtual interface, so it is more stable than a physical interface for RID use. A loopback address is configured by creating an interface and assigning an IP address.

```
Router(config)# interface loopback0
Router(config-if)# ip address 10.0.0.1 255.255.255.255
```

The loopback address does not need to be included in the OSPF routing process, but if you advertise it, you can ping or trace to it. This can help in troubleshooting.

A way to override the default RID selection is to statically assign it using the OSPF **router-id** command. Router ID is typically statically assigned for predictability should a process be forced to unexpectedly restart.

```
Router(config)# router ospf 8
Router(config-router)# router-id 10.0.0.1
```

Verify and Troubleshoot OSPF

The neighbor initialization process can be viewed using the **debug ip ospf adjacencies** command. The neighbor table can be seen with **show ip ospf neighbors**, which also identifies adjacency status and reveals the designated router and backup designated router. Use the **debug ip ospf packet** command to view all OSPF packets in real time.

Often, the first place OSPF issues are noticed is when inspecting the routing table: **show ip route**. To filter the routing table and show only the routes learned from OSPF, use **show ip route ospf**.

The command **show ip protocols** offers a wealth of information for any routing protocol issue. Use this command to verify parameters, timer values, identified networks, and OSPF neighbors (routing information sources).

Use **show ip ospf** to verify the RID, timers, and counters. Because wildcard masks sometimes incorrectly group interfaces to areas, another good place to check is **show ip ospf interface**. This shows the interfaces on which OSPF runs and their current correct assigned area.

OSPF Network Types

The SPF algorithm builds a directed graph—paths made up of a series of points connected by direct links. One of the consequences of this directed-graph approach is that the algorithm has no way to handle a multiaccess network, such as an Ethernet VLAN. The solution used by OSPF is to elect one router, called the Designated Router (DR), to represent the entire segment. Point-to-point links fit the SPF model perfectly and don't need any special modeling method. On a point-to-point link, no DR is elected, and all traffic is multicast to 224.0.0.5.

OSPF supports five network types:

- **NBMA:** Default for multipoint serial interfaces. RFC-compliant mode that uses DRs and requires manual neighbor configuration.
- **Point-to–multipoint (P2MP):** Doesn't use DRs so adjacencies increase logarithmically with routers. Resilient RFC-compliant mode that automatically discovers neighbors.
- **Point-to-multipoint nonbroadcast (P2MNB):** Proprietary mode that is used on Layer 2 facilities where dynamic neighbor discovery is not supported. Requires manual neighbor configuration.
- **Broadcast:** Default mode for LANs. Uses DRs and automatic neighbor discovery. Proprietary when used on WAN interface.
- **Point-to–point (P2P):** Proprietary mode that discovers neighbors and doesn't require a DR.

If the default interface type is unsatisfactory, you can statically configure it with the command **ip ospf network** under interface configuration mode:

```
Router(config-if)# ip ospf network point-to-multipoint
```

When using the NBMA or P2MP nonbroadcast mode, neighbors must be manually defined under the routing process:

```
Router(config-router)# neighbor 172.16.0.1
```

The command **show ip ospf interface** displays the network type for each link.

Designated Routers

On a multiaccess link, one of the routers is elected as a DR and another as a backup DR (BDR). All other routers on that link become adjacent only to the DR and BDR, not to each other. (They stop at the two-way state.) The DR is responsible for creating and flooding a network LSA (type 2) advertising the multiaccess link. NonDR (DROTHER) routers communicate with DRs using the IP address 224.0.0.6. The DRs use IP address 224.0.0.5 to pass information to other routers.

The DR and BDR are elected as follows:

Step 1. A router starting the OSPF process listens for OSPF Hellos. If none are heard within the dead time, it declares itself the DR.

Step 2. If Hellos from any other routers are heard, the router with the highest OSPF priority is elected DR, and the election process starts again for BDR. A priority of zero removes a router from the election.

Step 3. If two or more routers have the same OSPF priority, the router with the highest RID is elected DR, and the election process starts again for BDR.

After a DR is elected, elections do not take place again unless the DR or BDR are lost. Because of this, the DR is sometimes the first device that comes online with a nonzero priority.

The best way to control DR election is to set OSPF priority for the DR and BDR for other routers. The default priority is one. A priority of 0 means that a router cannot act as DR or BDR; it can be a DROTHER only. Priority can be set with the **ip ospf priority** command in interface configuration mode.

```
Router(config)# int fa 0/1
Router(config-if)# ip ospf priority 2
```

Nonbroadcast Multiaccess (NBMA) Networks

Routing protocols assume that multiaccess links support broadcast and have full-mesh connectivity from any device to any device. In terms of OSPF, this means the following:

- All Frame Relay or ATM maps should include the broadcast attribute.
- The DR and BDR should have full virtual circuit connectivity to all other devices.
- Hub-and-spoke environments should either configure the DR as the hub or use point-to-point subinterfaces, which require no DR.
- Partial-mesh environments should be configured using point-to-point subinterfaces, especially when no single device has full connectivity to all other devices. If there is a subset of the topology with full connectivity, that subset can use a multipoint subinterface.
- Full-mesh environments can be configured using the physical interface, but often logical interfaces are used to take advantage of the other benefits of subinterfaces.
- It might be necessary to statically identify neighbor IP addresses.

OSPF over Layer 2 and Layer 3 MPLS

Layer 2 and Layer 3 MPLS-based solutions were described in Chapter 2, "EIGRP." A Layer 2 connection uses EoMPLS, and OSPF operates just as it would on any other Ethernet network. It forms a neighbor relationship with the CE router across the WAN, and they elect a DR and BDR. The OSPF network type is Multiaccess Broadcast.

A Layer 3 MPLS VPN requires that the CE routers form an OSPF neighbor relationship with their connected PE router. The PE router appears to the enterprise as just another router within their network. The OSPF network type is determined by the type of link between the CE and PE. Carefully consider your area design when using this type of WAN.

Advanced OSPF Configuration

OSPF provides many different ways to customize its operation to fit your network needs. This section discusses route summarization, passive interfaces, default routes, stub areas, and virtual links.

OSPF Summarization

Summarization helps all routing protocols scale to larger networks, but OSPF especially benefits because its processes tax the memory and CPU resources of the routers. The SPF algorithm consumes all CPU resources when it runs. Summarization prevents topology changes from being passed outside an area and thus saves routers in other areas from having to run the SPF algorithm. OSPF's multiple databases use more memory the larger they are. Summarization decreases the number of routes exchanged, and thus the size of the databases. It localizes the impact of a topology change. OSPF can produce summaries within a classful network (VLSM) or summaries of blocks of classful networks (CIDR). There are two types of summarizations:

- **Inter-area (LSA type 3) route summarizations** are created on the ABR under the OSPF routing process using the **area range** command. A summary route will be advertised as long as at least one subnet within the summary is active in the area. The summary route's metric is the lowest cost route within the summary range. The router automatically creates a static route for the summary, pointing to Null0.

  The following command advertises 172.16.0.0/12 from area 1:

  ```
  Router(config-router)# area 1 range 172.16.0.0 255.240.0.0
  ```

- **External (LSA type 5) route summarization** is done on an ASBR using the **summary-address** command under the OSPF routing process. It can also be done on the ABR of a NSSA to summarize type 7 routes before advertising them as type 5. The router automatically creates a static route for the summary, pointing to Null0. The following example summarizes a range of external routes to 192.168.0.0/16 and injects a single type 5 route into OSPF.

  ```
  Router(config-router)#summary-address 192.168.0.0
    255.255.0.0
  ```

Passive Interface

The **passive-interface** command prevents OSPF from sending Hello messages out an interface. Thus an OSPF router does not discover neighbors or form an adjacency out that interface. To disable the protocol on one interface, use the routing protocol configuration command **passive-interface** *interface*. To turn off the protocol on all interfaces, use **passive-interface default**. You can then use **no passive-interface** *interface* for the ones that should run the protocol. See Chapter 2 for a configuration example.

OSPF Default Routes

The default route is a special type of summarization; it summarizes all networks down to one route announcement. This provides the ultimate benefit of summarization by reducing routing information to a minimum:

- Routers have a smaller routing table.
- Less use of router resources to advertise multiple routes.
- Routers do not need to keep information on external routes.

A default route is injected into OSPF as a type 5 route. There are several ways to use the router IOS to place a default route into OSPF. The best-known way is to use the **default-information** command under the OSPF routing process. This command, without the keyword **always**, advertises a default route learned from another source (such as a static route) into OSPF. If the **always** keyword is present, OSPF advertises a default even if that route does not already exist in the routing table. The **metric** keyword sets the starting metric for this route.

```
Router(config-router)# default-information originate [always]
[metric metric]
```

Alternatively, a default summary route can also be produced using the **summary-address** command or the **area range** command. These commands can cause the router to advertise a default route pointing to itself.

Reducing routing information in nonbackbone areas is a common requirement because these routers are typically the most vulnerable in terms of processor and speed, and the links that connect them usually have the least bandwidth. A specific concern is that an area will be overwhelmed by external routing information.

Stub and Not-So-Stubby Areas

Another way to reduce the route information advertised is to make an area a *stub* area. Configuring an area as a stub area forces its ABR to drop all external (type 5) routes and replaces them with a default route. To limit routing information even more, an area can be made *totally stubby* using the **no-summary** keyword on the ABR only. In that case, all interarea and external routes are dropped by the ABR and replaced by a default route. The default route starts with a cost of 1; to change it, use the **area default-cost** command. The example that follows shows area 2 configured as a totally stubby area, and the default route injected with a cost of 5:

```
Router(config-router)# area 2 stub no-summary
Router(config-router)# area 2 default-cost 5
```

Stub areas are attractive because of their low overhead. They do have some limitations, including the following:

- Stub areas can't include a virtual link.
- Stub areas can't include an ASBR.
- Stubbiness must be configured on all routers in the area.
- Area 0 cannot be a stub area.

Another kind of stub area is a *not-so-stubby area* (NSSA). NSSA is like a stub or totally stub area but enables an ASBR within the area. External routes are advertised as type 7 routes by the ASBR. The ABR converts them to type 5 external routes when it advertises them into adjacent areas. NSSA is configured with the **area nssa** command under the OSPF routing process. The **no-summary** keyword on the ABR configures the area as *totally NSSA*; this is a Cisco proprietary feature. By default, the ABR does not inject a default route back into an NSSA area. Use the **default-information-originate** keyword on the ABR or ASBR to create this route.

```
Router(config-router)# area 7 nssa [no-summary] [default-
 information-originate]
```

Virtual Links

OSPF requires that all areas be connected to area 0 and that area 0 must be contiguous. When this is not possible, you can use a virtual link to bridge across an intermediate area. Virtual links

- Connect areas that do not have a physical link to area 0. (This should be a temporary solution.)
- Connect a discontiguous area 0 (when merging two company networks, for instance. This should also be a temporary solution!)

Figure 3-3 shows a virtual link connecting two portions of the backbone area 0.

Area 1 is the transit area for the virtual link. Configure each end of a virtual link on the ABRs of the transit area with the command **area** *area-number* **virtual-link** *router-id*. Each end of the link is identified by its RID. The area listed in the command is the transit area, not the area being joined by the link. The configuration for R1 is

```
R1(config)# router ospf 1
R1(config-router)# area 1 virtual-link 10.20.20.20
```

Figure 3-3 OSPF Virtual Link

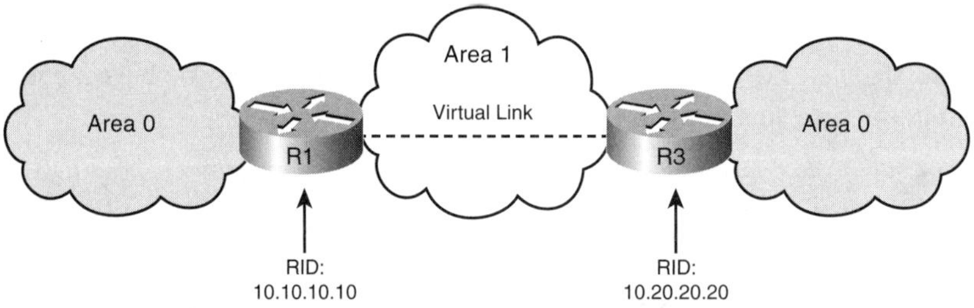

The configuration for R3 is

```
R3(config)# router ospf 1
R3(config-router)# area 1 virtual-link 10.10.10.10
```

Verify that the virtual link is up with the **show ip ospf virtual-links** command. Additionally, virtual interfaces are treated as actual interfaces by the OSPF process, and thus, their status can be verified with the **show ip ospf interface** *interface-id* command.

OSPF Authentication

For security purposes, you can configure OSPF to authenticate every OSPF packet and the source of every OSPF routing update. By default, the router does no authentication. OSPF supports two types of authentication:

- Simple (plain text) authentication
- MD5 authentication

The following example shows a router configured for simple password authentication in OSPF area 1, using a password (or *key*) of "simple." Note that authentication commands are necessary both under the OSPF process and the interface configuration. All OSPF neighbors reachable through an interface configured for authentication must use the same password. You can, however, use different passwords for different interfaces.

```
Router(config)# int gi0/0
Router(config-if)# ip ospf authentication-key simple
Router(config-if)# ip ospf authentication
Router(config-if)# !
Router(config-if)# router ospf 1
Router(config-router)# area 1 authentication
```

The next example shows the same router configured for OSPF MD5 authentication for area 0, using a password of "secure." Note that the commands are slightly different. The optional keyword **message-digest** is required in two of the commands, and a key number must be specified. Any neighbors reachable through the Gi0/1 interface must also be configured with the same key.

```
Router(config-router)# int gi0/1
Router(config-if)# ip ospf message-digest-key 2 md5 secure
Router(config-if)# ip ospf authentication message-digest
Router(config-if)# !
Router(config-if)# router ospf 1
Router(config-router)# area 0 authentication message-digest
```

Use the following commands to verify and troubleshoot OSPF authentication:

- **debug ip ospf adj:** The debug shows an error message if there is a key mismatch.
- **show ip ospf neighbor:** If a neighbor relationship has been established, you can assume the authentication worked properly.
- **show ip route:** Verify that route information is being exchanged between the two authenticating routers.

CHAPTER 4

Optimizing Routing

There are times when you need to go beyond just turning on a routing protocol in your network. You might need to control exactly which routes are advertised or redistributed, or which paths are chosen. You might also need to use multiple routing protocols. Network performance can suffer when routing is not optimized. Excessive routing updates lead to extra CPU usage because of the amount of routing information and the frequency of updates. Running multiple protocols requires extra router resources and might result in suboptimal paths. Incorrectly configured route filters can lead to routing issues.

Controlling Routing Updates

Cisco IOS provides several ways to control routing updates:

- Route Maps
- Prefix Lists
- Distribute Lists
- Passive Interface

When a route update arrives at a router's interface, the router checks to see if a route filter is associated with that interface. If not, the update processes normally. If there is a filter, the router checks for an entry matching the update. If there is no matching entry, the update is dropped. If a matching entry exists, the router processes the update based on instructions in the filter.

Route Maps

Route maps are a bit like programs that use an if/then/else decision-making capability. They *match* traffic against certain conditions and then set specified options for that traffic. Each statement has a sequence number, statements are read from the lowest number to highest, and the router stops reading when it gets a match. The sequence number can be used to insert or delete statements. Like an access list, there is an implicit "deny" at the end of each route map; any traffic not matched with a route map statement is denied. Some uses for route maps include

- **Filtering redistributed routes:** Use the **route-map** keyword in the redistribute command.
- **Policy-based routing:** To specify which traffic should be policy routed, based on very granular controls.
- **BGP policy:** To control routing updates and to manipulate path attributes.

Route Map Syntax

Route maps are created with the global command:

```
Router(config)# route-map {tag} permit | deny [sequence_number]
```

Each statement in a route map begins this same way, with the same route map name but different sequence numbers, and with match and set conditions below it. *Permit* means that any traffic matching the match conditions is processed by the route map statement. *Deny* means that any traffic matching the match conditions is not processed by the route map statement.

Route Map Match and Set Conditions

Each route map statement can have from none to multiple **match** and **set** conditions. If no **match** condition exists, the statement matches anything, similar to a "permit any" in an access list. If there is no **set** condition, the matching traffic is either permitted or denied, with no other conditions being set.

Multiple match conditions on the same line use a logical OR. For example, the router interprets **match a b c** as "**match a** or **b** or **c**." Multiple match conditions on different lines use a logical AND. For example, the router interprets the following route map statement as "**match a** and **b** and **c**":

```
route-map Logical-AND permit 10
 match a
 match b
 match c
```

In route redistribution, some common conditions to **match** include

- **ip address:** Refers the router to an access list that permits or denies networks.
- **ip address prefix-list:** Refers the router to a prefix-list that permits or denies IP prefixes.

- **ip next-hop:** Refers the router to an access list that permits or denies next-hop IP addresses.
- **ip route-source:** Refers the router to an access list that permits or denies advertising router IP addresses.
- **length:** Permits or denies packets based on their length in bytes.
- **metric:** Permits or denies routes with the specified metric from being redistributed.
- **route-type:** Permits or denies redistribution of the route type listed, such as internal or external.
- **tag:** Routes can be labeled (tagged) with a number, and route maps can look for that number.

In route redistribution, some common conditions to set are

- **metric:** Sets the metric for redistributed routes.
- **metric-type:** Sets the type, such as E1 for OSPF.
- **tag:** Tags a route with a number that can be matched on later by other route maps.

Controlling Route Redistribution Using Route Maps

The following configuration example shows a route map named BGP-LP with three statements that control which routes will be redistributed from OSPF into BGP. The router has already been configured with two access lists, numbered 23 and 103 (not shown.) The first route map statement, with sequence number 10, is a *permit* statement. The **match** condition tells it to use access list 23. Any traffic permitted by access list 23 matches this statement and will be redistributed into BGP. Any traffic explicitly denied by access list 23 will not be redistributed into BGP. The **set** condition tells it to set a BGP local preference for all traffic that matches statement 10. Traffic not matching access list 23 will be checked against the second route map statement.

The second route map statement, sequence number 20, is a *deny* statement that matches access list 103. Any traffic permitted by access list 103 will be denied by this statement and thus will not be redistributed. Any traffic explicitly denied by access list 103 will be ignored by this statement and checked against the next route map statement. This route map statement has no **set** conditions. Traffic not matching route map statements 10 or 20 will be checked against statement 30.

The third route map statement, sequence number 30, is a *permit* statement with no **match** or **set** conditions. This statement matches everything and sets nothing, thus permitting all other traffic without changing it. Without this statement, all other traffic would be denied.

Lastly, the route map is applied to the redistribution command to filter routes redistributed from OSPF into BGP:

```
Router(config)# route-map BGP-LP permit 10
Router(config-route-map)# match ip address 23
Router(config-route-map)# set local-preference 200
Router(config-route-map)# !
Router(config-route-map)# route-map BGP-LP deny 20
Router(config-route-map)# match ip address 103
Router(config-route-map)# !
Router(config-route-map)# route-map BGP-LP permit 30
!
Router(config)# router bgp 65001
Router(config-router)# redistribute ospf 1 route-map BGP-LP
```

Policy-Based Routing Using Route Maps

Policy-based routing overrides the normal routing process. Normal routing is done based on the destination IP address. Policy-based routing is based on source IP address or interface, or packet length. Create a route map statement that matches an access list, a specific IP address, or a packet length range. Then set either a next-hop IP address or an outbound interface for any traffic that matches the statement. Next, apply the route map either to an inbound interface or to the router itself to control locally generated traffic. The following configuration example shows a route map named LOCAL that matches the source addresses in access list 1. It assigns a next-hop IP address of 10.1.1.1 to this traffic. Because it is applied to the local router, it will be used only for traffic generated by the router itself.

```
Router(config)# route-map LOCAL
Router(config-route-map)# match ip address 1
Router(config-route-map)# set ip next-hop 10.1.1.1
!
Router(config)# ip local policy route-map LOCAL
```

Route map INT, shown in the following example, has no match condition and thus matches all traffic. It sets an outbound interface. Because it is applied to an interface, its policy routes all inbound traffic from that interface:

```
Router(config)# route-map INT
Router(config-route-map)# set interface fa0/1
!
```

```
Router(config)# int fa0/0
Router(config-if)# ip policy route-map INT
```

Verify policy routing with the **debug ip policy** command. See Chapter 5, "Path Control," for more information on policy-based routing.

Tagging Routes Using a Route Map

Another use for a route map is to tag routes as they are redistributed from one protocol to another. Then you can deny tagged routes from being redistributed back into the original protocol. For instance, supposed you are mutually redistributing routes between OSPF and EIGRP. You can tag EIGRP routes as you redistribute them into OSPF. Then when you redistribute OSPF routes back into EIGRP, you can deny those tagged routes. The following example illustrates this.

```
Router(config)# route-map EIGRP2OSPF deny 5
Router(config-route-map)# match tag 1
Router(config-route-map)# route-map EIGRP2OSPF permit 10
Router(config-route-map)# set tag 2
!
Router(config)# route-map OSPF2EIGRP deny 5
Router(config-route-map)# match tag 2
Router(config-route-map)# route-map OSPF2EIGRP permit 10
Router(config-route-map)# set tag 1
!
Router(config)# router eigrp 1
Router(config-router)# redistribute ospf 2 route-map OSPF2EIGRP
 metric 1 1 1 1 1500
!
Router(config-router)# router ospf 2
Router(config-router)# redistribute eigrp 1 route-map EIGRP2OSPF
 subnets
```

Prefix Lists

A prefix list matches both the subnet, or *prefix*, and the number of bits in the subnet mask. Similar to an access list, it consists of one or more statements permitting or denying prefixes. Routers evaluate the prefix statements in order, stopping if they find a match. There is an implicit "deny all" at the end of the prefix list. The command syntax follows:

```
ip prefix-list {list-name [seq number] {deny | permit}
 network/length [ge ge-length] [le le-length]
```

The meaning of each command field is detailed in Table 4-1.

Table 4-1 The ip prefix-list Command

Command Field	Meaning
`list-name`	Gives a name to the prefix list. Prefix lists are named, not numbered.
`seq number`	[Optional] Assigns a sequence number to the prefix list statement. Statements are numbered in increments of 5 by default, enabling a statement to be inserted between two others by using the **seq** option.
`deny | permit`	Denies or permits the matching prefix.
`network/length`	Configures the prefix and number of bits that must be matched. If no ge or le option is given, the length also equals the length of the subnet mask.
`ge` *`ge-length`*	[Optional] Stands for "greater than or equal to." Specifies the minimum number of bits a subnet mask must have to match the statement.
`le` *`le-length`*	[Optional] Stands for "lesser than or equal to." Specifies the maximum number of bit a subnet mask can have to match the statement.

Some sample prefix lists include

- **ip prefix-list CCNP permit 0.0.0.0/0:** Permits only a default route.
- **ip prefix-list CCNP permit 0.0.0.0/0 le 32:** Permits all routes (equivalent to a "permit any" in an access list.) The prefix 0.0.0.0/0 means that none of the prefix bits must be matched. "Le 32" means that the subnet mask must be less than or equal to 32. Thus any network will match this statement.
- **ip prefix-list CCNP permit 0.0.0.0/0 ge 32:** Permits only host routes. The prefix 0.0.0.0/0 means that none of the prefix bits must be matched. "Ge 32" means that the subnet mask must be exactly 32 bits, thus this statement matches only host routes.
- **ip prefix-list CCNP permit 10.0.0.0/8 ge 24 le 24:** Permits any route whose first 8 bits equal 10, with a subnet mask of exactly 24 bits.

Before taking the ROUTE exam, be sure you understand and can interpret prefix lists.

Prefix lists can be used in a route map to control redistribution of networks. They can also be applied to a BGP neighbor to filter routing updates to that neighbor.

Distribute Lists

A distribute list enables you to filter both routing updates and routes being redistributed, through the use of an access list. Configure an access list that permits the routes to be advertised or redistributed, and then link that access list to the routing process with the **distribute-list** command, given under router configuration mode. This command has two options:

- **distribute-list** *access-list* **in**–Filters updates as they come in an interface. For OSPF, this controls routes placed in the routing table but not the database. For other protocols, this controls the routes the protocol knows about.
- **distribute-list** *access-list* **out**–Filters updates going out of an interface and also updates being redistributed out of another routing protocol into this one.

Passive Interfaces

The **passive-interface** command is another way to control routing updates because it prevents any updates from sending out an interface that is marked as passive. OSPF and EIGRP do not send Hello messages out a passive interface, and thus do not discover any neighbors. RIP does not send updates out a passive interface but listens for inbound updates. The EIGRP and OSPF chapters have a more in-depth description of this command.

Using Multiple Routing Protocols

There are several reasons you might need to run multiple routing protocols in your network. Some include

- Migrating from one routing protocol to another, where both protocols will run in the network temporarily
- Applications that run under certain routing protocols but not others
- Areas of the network under different administrative control (Layer 8 issues)
- A multivendor environment in which some parts of the network require a standards-based protocol

Configuring Route Redistribution

Route redistribution is used when routing information must be exchanged among the different protocols or routing domains. Only routes that are in the routing table and learned via the specified protocol are redistributed. Each protocol has some unique characteristics when redistributing, as shown in Table 4-2.

Table 4-2 Route Redistribution Characteristics

Protocol	Redistribution Characteristics
`RIP`	Default metric is Infinity. Metric must be set, except when redistributing static or connected routes, which have a metric of 1.
`OSPF`	Default metric is 20. Can specify the metric type; the default is E2. Must use **subnets** keyword or only classful networks are redistributed.
`EIGRP`	Default metric is Infinity. Metric must be set, except when redistributing static or connected routes, which get their metric from the interface. Metric value is "bandwidth, delay, reliability, load, MTU." Redistributed routes have a higher administrative distance than internal ones.
`Static/Connected`	To include local networks not running the routing protocol, you must redistribute connected interfaces. You can also redistribute static routes into a dynamic protocol.
`BGP`	Metric (MED) is set to IGP metric value.

You can redistribute only between protocols that use the same protocol stack, such as IP protocols, which cannot advertise IPX routes. To configure redistribution, issue this command under the routing process that is to receive the new routes:

```
Router(config-router)# redistribute {route-source} [metric metric]
  [route-map tag]
```

Seed Metric

Redistribution involves configuring a routing protocol to advertise routes learned by another routing process. Normally, protocols base their metric on an interface value, such as bandwidth, but a redistributed route is not associated with an interface. Protocols use incompatible metrics, so the redistributed routes must be assigned a new metric compatible with the new protocol.

A route's starting metric is called its *seed metric*. Set the seed metric for all redistributed routes with the **default-metric** [*metric*] command under the routing process. To set the metric for specific routes, either use the **metric** keyword when redistributing or use the **route-map** keyword to link a route map to the redistribution. After the seed metric is specified, it increments normally as the route is advertised through the network (except for certain OSPF routes).

Administrative Distance

When a router receives routes to the same destination network from more than one routing process, it decides which to put in the routing table by looking at the administrative distance (AD) value assigned to the routing process. The route with the lowest AD is chosen. Table 4-3 shows administrative distance values.

Table 4-3 Administrative Distance

Routing Information Source	Administrative Distance
Connected interface	0
Static route	1
EIGRP summarized route	5
BGP external route	20
EIGRP internal route	90
IGRP	100
OSPF	110
IS-IS	115
RIP	120
EIGRP external route	170
BGP internal route	200
Unknown	255

AD can be changed for all routes of a process or only for specific routes within a process. The command for all IGPs except EIGRP is

```
Router(config-router)# distance administrative_distance {address
 wildcard-mask} [access-list-number | name]
```

Using the **address/mask** keywords in the command changes the AD of routes learned from the neighbor with that IP address. An entry of **0.0.0.0**

255.255.255.255 changes the AD of all routes. Specifying an access list number or name changes the AD only on networks permitted in the ACL.

EIGRP and BGP have different AD values for internal and external routes, so you have to list those separately when using the command with those protocols. BGP also enables you to change the AD for locally generated routes. For these protocols, the commands are

```
Router(config-router)# distance eigrp internal-distance external-
  distance
Router(config-router)# distance bgp external-distance internal-
  distance local-distance
```

Route redistribution can cause suboptimal routing; one way to correct this is to adjust AD. Figure 4-1 shows a network with two routing domains: RIP and OSPF.

Figure 4-1 Controlling Routing with AD

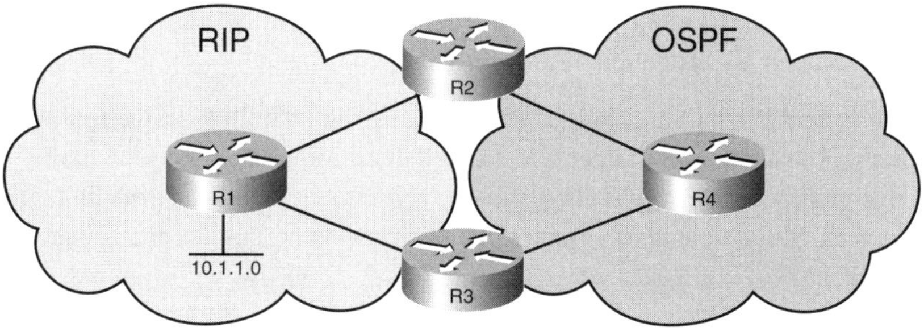

R2 redistributes its RIP routes into OSPF. These routes inherit OSPF's AD when they are advertised to R4, which then advertises them to R3 as OSPF routes.

R3 now knows about the 10.1.1.0 network from two routing processes: RIP, with an AD of 120, and OSPF, with an AD of 110. The shortest path is the RIP route through R1. The OSPF path goes through R4 and R2, and then to R1—a much longer path. But, based on AD, R3 puts the OSPF path in its routing table.

To prevent this, increase the AD of the redistributed RIP routes when OSPF advertises them. Note that this doesn't change all OSPF routes, just the ones learned from RIP. The commands given on R2 (the router doing the initial redistribution) are shown here:

```
Router(config)# access-list 10 permit 10.1.1.0
!
Router(config)# router ospf 1
```

```
Router(config-router)# redistribute rip subnets
Router(config-router)# distance 125 0.0.0.0 255.255.255.255 10
```

The AD is increased to 125 for routes from all neighbors, if they match the network permitted in access list 10. Now R3 hears about the 10.1.1.0 network from RIP with an AD of 120, and from OSPF with an AD of 125. The RIP route is put into the routing table based on its lower AD.

Routing protocols that assign a higher AD to external routes, EIGRP and BGP, accomplish a similar result automatically. OSPF can be configured to do so with the **distance ospf external** command.

Planning Route Redistribution

Plan carefully before redistributing routes between protocols. Different protocols have incompatible routing information and different convergence times. First, decide which is the core, or main, protocol and which is the edge protocol. Decide if you will do one-way or two-way, and single point or multipoint redistribution.

One-way redistribution involves redistributing routes from the edge routing protocol into the core protocol. Static or default routes must be used in the edge protocol. Two-way redistribution involves redistributing routes mutually between both core and edge protocols. No static routes are needed because both protocols know all routing information.

One-way and two-way redistribution at just one router within the network is considered safe because traffic between administrative domains has only one exit point, thus routing loops are not a problem. Redistribution at multiple routers within the network can cause routing loops and suboptimal routing.

With multipoint one-way redistribution:

- Use a routing protocol that uses different ADs for external and internal routes (EIGRP, OSPF, and BGP).
- Ensure that the AD of the redistributed external routes is higher than the AD of the protocol where they originated.

Multipoint two-way redistribution adds the following considerations:

- Ensure that only internal routes are redistributed from each protocol. You can do this by tagging the routes and then filtering based on tags when redistributing.
- Adjust the metric of the redistributed routes.

- Consider using a default route to avoid multipoint two-way redistribution.

Redistribution Techniques

Try to design your route redistribution as safely as possible. The options include

- Redistribute all edge information into the core, but send only a default route into the edge.
- Redistribute all edge information into the core, but redistribute multiple static routes into the edge.
- Redistribute routes in both directions, but filter to prevent routes from being redistributed back into their original administrative domain.
- Redistribute all routes in both directions, but increase the AD for external routes.

Redistribution Notes

The IPv6 commands to redistribute routes between protocols or between multiple instances of a protocol are just like the ones in IPv4. Under the routing protocol configuration mode, issue the command **redistribute** *route-source* and specify any options such as a route map if desired.

Some points to remember about redistributing routes follow:

- A router redistributes only routes learned by the source protocol. For instance, if you redistribute connected routes into the protocol, it will advertise them but not redistribute them.
- When redistributing routes into BGP, you can use the keyword **include-connected** to get the connected routes into BGP.
- When you redistribute routes between two OSPF processes, the routes are advertised into the new process as Type 2.
- You generally want to include the **subnets** keyword on routes distributed from another routing protocol into OSPF. Otherwise, only routes that use their default classful subnet mask are redistributed.
- Be sure to specify a seed metric when redistributing routes into RIP. Otherwise the routes start with a metric of 16, which RIP interprets as "unreachable."

- If you redistribute in multiple places, check the path that traffic takes. You might run into suboptimal routing. A way to fix this is to tune the administrative distance for some of the routes.
- BGP does not redistribute routes learned via IBGP into an IGP by default. To change this behavior, use the router configuration command **bgp redistribute-internal**.

CHAPTER 5

Path Control

In general, link redundancy is a good thing, but it can also lead to some network problems. You might want to manually control the route taken by some or all of your traffic to provide a predictable and deterministic traffic flow. Path control can prevent suboptimal routing, ensure path availability, provide optimized performance for specific applications, and provide load sharing among various paths.

A good path control strategy understands that traffic is bidirectional and considers both inbound and outbound traffic. Asymmetric routing—where traffic exits via one link and enters via another—is not inherently bad. But you might need to minimize it when using stateful devices such as firewalls, or with sensitive applications such as voice. A good routing plan requires a good design. Design IP addresses that can be summarized and use redistribution and passive interfaces appropriately.

The previous chapter covered several ways to influence paths: route maps, prefix lists, distribute lists, administrative distance, and route tags. This chapter covers some others: Offset-lists, IP SLA, Policy-based Routing (PBR), Optimized Edge Routing, and Virtual Routing and Forwarding (VRF).

Using Offset-lists

An offset-list is a way to increase the metric of routes. You might do this to cause a router to choose what it would normally consider a less desirable path, or to load balance over paths that would normally have unequal methods. The command uses an access list, so you can also use this command to send a subset of traffic over a different path. Historically, offset-lists were used with RIP because it only looked at hop count, and thus might choose a slow path that had a lower hop count than a fast path. Only RIP and EIGRP support offset-lists.

Configure an offset-list with the command **offset-list** {*access-list-number* | *access-list-name*} {**in** | **out**} *offset* [*interface-type interface-number*]. This command is given in router configuration mode. The offset amount is added to the hop count in RIP and is added to the delay value in EIGRP. Be careful when changing an EIGRP metric value and test thoroughly. The following example shows a delay value of 2000 added to the EIGRP metric of the two routes permitted in access list Offset, when they are advertised in interface FastEthernet 0/0.

```
R1(config)# ip access-list standard Offset
R1(config-std-nacl)# permit 172.31.1.0 0.0.0.255
R1(config-std-nacl)# permit 172.16.1.0 0.0.0.255
!
R1(config)# router eigrp 100
R1(config-router)# offset-list Offset in 2000 fa0/0
```

You can configure more than one offset-list. If you specify an interface, the offset-list is considered an extended offset-list and has precedence over a normal offset-list.

Using IOS IP SLA

IP SLA is a feature that enables a Cisco router or switch to simulate specific types of traffic and send it either to an IP address or to a receiver, called a "responder." IP SLA probes can simulate various types of traffic, such as HTTP, FTP, DHCP, UDP jitter, UDP echo, HTTP, TCP connect, ICMP echo, ICMP path echo, ICMP path jitter, and DNS, and can report statistics such as path jitter. It has highly granular application configuration options such as TCP/UDP port numbers, TOS byte, and IP prefix bits. With IP SLA you can measure network performance and host reachability. This is useful for path control because it enables you to switch to a backup path if network performance on the primary path degrades, or if a link failure occurs someplace in the primary path.

To use IP SLA for path control, you must

1. Create a monitor session on the probe source device.
2. Define the probe by specifying traffic type, destination IP address, and any other desired variables such as DSCP value.
3. Schedule the probe beginning and ending times.
4. Define a tracking object that is linked to the monitor session.
5. Link the tracking object to a static route.

The destination can be any trusted device that responds to the traffic you send. To use IP SLA for bringing up a backup link, choose a destination that will reflect problems in the ISP's network. One benefit of using a Cisco device as the responder is that it can add time stamps to help measure latency and jitter. These time stamps take into account the device processing time so that the measurement reflects only network latency. The configuration of a Cisco responder is simple. Use the global command **ip sla responder.**

Figure 5-1 shows a network with two connections to the Internet, one primary and one backup. The edge router has a static default route pointing to each provider.

Figure 5-1 Using IP SLA for Path Control

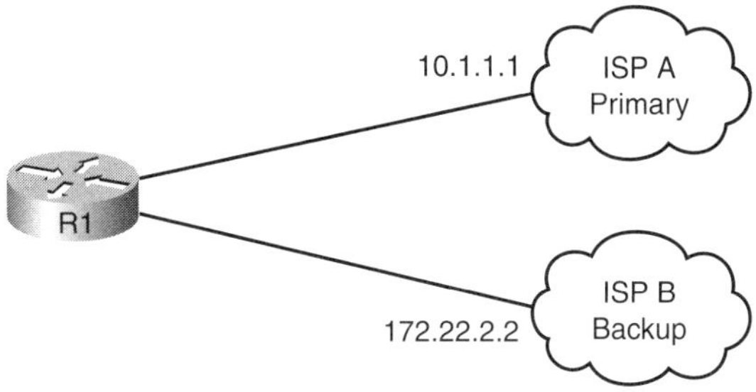

In the following example, router R1 is configured to conditionally announce a default route based on the IP SLA probe response. Two IP SLA monitor sessions are configured to send a ping every 10 seconds to a DNS server within each ISP's network. A separate tracking object tracks reachability to each server. The two default route statements tell the router to give the primary route an administrative distance of 2 if the tracked object is reachable. The backup route is assigned an administrative distance of 3 if its tracked object is reachable.

```
R1(config)# ip sla 1
R1(config-ip-sla)# icmp-echo 10.1.1.50
R1(config-ip-sla-echo)# frequency 10
!
R1(config)# ip sla 2
R1(config-ip-sla)# icmp-echo 171.22.2.52
R1(config-ip-sla-echo)# frequency 10
!
R1(config)# ip sla schedule 1 life forever start-time now
R1(config)# ip sla schedule 2 life forever start-time now
!
R1(config)# track 1 ip sla 1 reachability
R1(config)# track 2 ip sla 2 reachability
!
!Primary route
R1(config)# ip route 0.0.0.0 0.0.0.0 10.1.1.1 2 track 1
```

```
!Backup route
R1(config)# ip route 0.0.0.0 0.0.0.0 172.22.2.2 3 track 2
```

Under normal circumstances, the default route with an AD of 2 would be installed in the IP routing table instead of the one with an AD of 3. But when the primary DNS server is not reachable, the primary route is withdrawn, and the backup route with an AD of 3 is installed in the routing table.

Additionally, you can combine IP SLA tracking with policy-based routing to provide redundancy for specific traffic types on a per-interface basis. For instance, after an access list permitting the interesting traffic is specified, a route-map such as the following can be configured. In the example route-map, x.x.x.x and y.y.y.y represent different next hops to send the traffic specified in ACL 101. The **track 1** and **track 2** keywords tie the configuration back to the SLA groups configured in the previous example. You then apply the policy map under the incoming interfaces just as with normal policy routing.

```
route-map REDUNDANT permit 10
 match ip address 101
 set ip next-hop verify-availability x.x.x.x 10 track 1
 set ip next hop verify-availability y.y.y.y 20 track 2
!

interface FastEthernet0/1
 description LAN Interface
 ip policy route-map REDUNDANT
```

Policy-Based Routing

Normal IP routing chooses a path based on the destination IP address. Policy-based routing (PBR) lets you route traffic based on other variables. It uses a route-map to match traffic and then sets either a next-hop address or an exit interface. It can also mark the traffic that it policy routes. Any traffic not matched in the route-map is routed normally. Policy-based routing can be applied both to traffic entering the router and to traffic originated by the router.

Some benefits of policy routing include

- Ability to route based on traffic source, and other attributes
- Ability to set QoS markings
- Ability to force load sharing between unequal paths
- Ability to allocate traffic among multiple paths based on traffic attributes

Use the following steps to configure policy-based routing:

1. Configure a route-map that matches the desired traffic and uses the **set** command to define the actions for that traffic.
2. Optionally enable fast-switched PBR. CEF-switched PBR is enabled automatically whenever CEF is enabled.
3. Apply the route map either to an incoming interface or to traffic generated by the router.
4. Verify the configuration.

Some typical attributes to match in the route map include source or destination address with an access list and packet length. If no match criteria are specified, all packets are considered a match.

You can choose to set IP precedence, but the most typical setting determines how the traffic leaves the router. There are four ways to do this. If there are multiple **set** statements, the router evaluates them in this order:

1. **set ip next-hop** *ip-address*: When this command is given, the router checks to see if the next-hop address is reachable. If so, it forwards the traffic toward that address. If not, it uses the routing table.
2. **set interface** *interface-type interface-number:* Multiple interfaces can be listed. When this command is given, the router checks that it has an explicit route for the destination network in its routing table before forwarding the traffic out the specified interface. If it does not, this command is ignored. A default route is not considered an explicit route. Listing multiple interfaces under the set command allows for redundancy if the first interface fails or goes down. The router uses the first active interface listed.
3. **set ip default next-hop** *ip-address*: If the routing table contains an explicit route for the destination network, that route is used and this command is ignored. If no explicit route exists, this command is executed. A default route is not considered an explicit route.
4. **set default interface** *interface-type interface-number*: If the routing table contains an explicit route for the destination network, that route is used and this command is ignored. If no explicit route exists, traffic is forwarded out the specified interface. A default route is not considered an explicit route.

To apply the PBR route-map to an interface, use the interface command **ip policy route-map** *name*. To apply PBR to packets originated by the router itself, use the global command **ip local policy route-map** *name*. The following example shows a PBR route map that matches traffic in access list 101, and policy routes it to a next hop of 10.1.1.1. The policy is applied to interface fa0/0, so all traffic entering that interface is evaluated against the route-map. Packets not permitted by access list 101 are routed normally (destination routed.)

```
route-map Policy permit 10
 match ip address 101
 set ip next hop 10.1.1.1
!
interface FastEthernet0/0
 ip policy route-map Policy
```

PBR can also use IP SLA tracking, described previously in the IP SLA section. Verify your configuration with the commands **show ip policy** and **show route-map** {*name*}.

OER and VRF

With OER, Border Routers monitor WAN performance and report the information to a Master Controller router. If WAN performance falls within configured ranges, no change is made to the default routing. If performance begins to degrade for a specific link or network, the Controller notifies its Border Routers to reroute traffic, perhaps by adding a static route or changing routing protocol parameters.

VRFs are a way to segment traffic. You might think of them as Layer 3 VLANs. Just as VLANs create virtual switches with segregated CAM tables, VRFs create virtual routers with segregated routing tables. This lets you separate guest traffic from employee traffic, for instance. Traffic for different VRFs can be routed over different paths.

CHAPTER 6

BGP and Internet Connectivity

Planning an Internet Connection

Consider your company's needs when planning your Internet connection. If all you need is one-way connectivity to enable internal users to connect to sites on the Internet, a private IP address space with Network Address Translation (NAT) should suffice. If external users need to connect to resources, such as servers, inside your network, you need some public IP addresses. You might combine these with private addresses and NAT for your users.

If external users must connect to your internal resources, you should plan the following:

- How many public IP addresses will you need?
- Should you get your IP addresses from your ISP or acquire your own? If you elect to acquire your own addresses, you also need a public Autonomous System (AS) number.
- What link type and speed will you need to support all the external connections plus your internal users?
- Will you use static or dynamic routing?
- How much redundancy will you need? This includes link redundancy and ISP redundancy.

To Route or Not to Route?

If your ISP connection is a Layer 2 circuit emulation, there is no need to run a routing protocol with the ISP.

If you use MPLS VPNs, you either use static routes or run a dynamic routing protocol with the ISP edge router. This might be either one of the IGPs (EIGRP, OSPF, or RIP) or Border Gateway Protocol (BGP).

If you need only a default route pointing to your ISP, static routes work. The provider needs to create static routes pointing to your network and redistribute them into its routing protocol.

BGP is a good choice if you connect to multiple ISPs, you need to control how traffic enters or exits your company, or you need to react to Internet topology changes.

BGP Route Options

You have a choice of three ways to receive BGP routes from an ISP:

- **Default routes from each provider:** This is simple to configure and results in low use of bandwidth and router resources. The internal network's IGP metric determines the exit router for all traffic bound outside the autonomous system. No BGP path manipulation is possible, so this can lead to suboptimal routing if you use more than one ISP.
- **Default routes plus some more specific routes:** This option results in medium use of bandwidth and router resources. It enables you to manipulate the exit path for specific routes using BGP so that traffic takes a shorter path to networks in each ISP. Thus path selection is more predictable. The IGP metric chooses the exit path for default routes.
- **All routes from all providers:** This requires the highest use of bandwidth and router resources. It is typically done by large enterprises and ISPs. Path selection for all external routes can be controlled via BGP policy routing tools.

Types of ISP Connections

A site with a single ISP connection is *single-homed.* This is fine for a site that does not depend heavily on Internet or WAN connectivity. Either use static routes, or advertise the site routes to the ISP and receive a default route from the ISP.

A *dual-homed* site has two connections to the same ISP, either from one router or two routers. One link might be primary and the other backup, or the site might load balance over both links. Either static or dynamic routing would work in this case.

Multihoming means connecting to more than one ISP at the same time. It is done for redundancy and backup if one ISP fails, and for better performance if one ISP provides a better path to frequently used networks. This also gives you an ISP-independent solution. BGP is typically used with multihomed connections.

You can take multihoming a step further and be *dual-multihomed*, with two connections to multiple ISPs. This gives the most redundancy. BGP is used with the ISPs and can be used internally also.

BGP Overview

BGP is an external gateway protocol, meant to be used between different networks. It is the protocol used between Internet service providers (ISPs) and also can be used between an Enterprise and an ISP. BGP was built for reliability, scalability, and control, not speed. Because of this, it behaves differently from the protocols covered so far in this book:

- BGP stands for Border Gateway Protocol. Routers running BGP are termed *BGP speakers*.
- BGP uses the concept of autonomous systems (AS). An *autonomous system* is a group of networks under a common administration. The Internet Assigned Numbers Authority (IANA) assigns AS numbers: 1 to 64511 are public AS numbers and 64512 to 65535 are private AS numbers.
- Autonomous systems run Interior Gateway Protocols (IGP) within the system. They run an Exterior Gateway Protocol (EGP) between them. BGP version 4 is the only EGP currently in use.
- Routing between autonomous systems is called *interdomain routing*.
- The administrative distance for EBGP routes is 20. The administrative distance for IBGP routes is 200.
- BGP neighbors are called peers and must be statically configured.
- BGP uses TCP port 179. BGP peers exchange incremental, triggered route updates and periodic keepalives.
- Routers can run only one instance of BGP at a time.
- BGP is a path-vector protocol. Its route to a network consists of a list of autonomous systems on the path to that network.
- BGP's loop prevention mechanism is an autonomous system number.

> When an update about a network leaves an autonomous system, that autonomous system's number is prepended to the list of autonomous systems that have handled that update. When an autonomous system receives an update, it examines the autonomous system list. If it finds its own autonomous system number in that list, the update is discarded.

In Figure 6-1, BGP routers in AS 65100 see network 10.1.1.0 as having an autonomous system path of 65200 65300 65400.

Figure 6-1 BGP AS-Path Advertisement

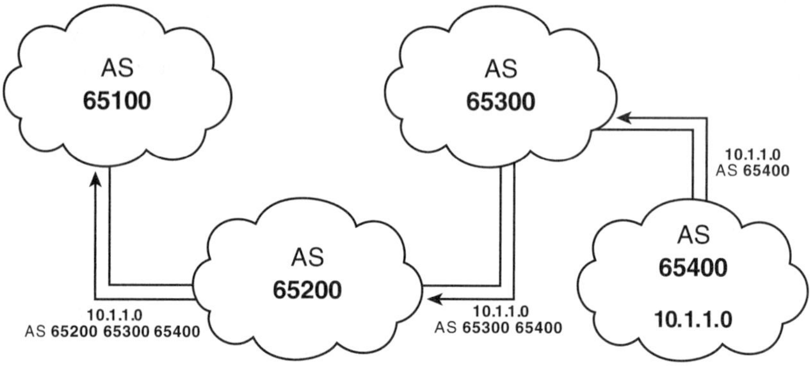

Use BGP when the AS is multihomed, when route path manipulation is needed, or when the AS is a transit AS. (Traffic flows through it to another AS, such as with an ISP.)

Do not use BGP in a single-homed AS, with a router that does not have sufficient resources to handle it, or with a staff that does not have a good understanding of BGP path selection and manipulation.

BGP Databases

BGP uses three databases. The first two listed are BGP-specific; the third is shared by all routing processes on the router:

- **Neighbor database:** A list of all configured BGP neighbors. To view it, use the **show ip bgp summary** command.
- **BGP database, or RIB (Routing Information Base):** A list of networks known by BGP, along with their paths and attributes. To view it, use the **show ip bgp** command.

- **Routing table:** A list of the paths to each network used by the router, and the next hop for each network. To view it, use the **show ip route** command.

BGP Message Types

BGP has four types of messages:

- **Open:** After a neighbor is configured, BGP sends an open message to try to establish peering with that neighbor. Includes information such as autonomous system number, router ID, and hold time.
- **Update:** Message used to transfer routing information between peers. Includes new routes, withdrawn routes, and path attributes.
- **Keepalive:** BGP peers exchange keepalive messages every 60 seconds by default. These keep the peering session active.
- **Notification:** When a problem occurs that causes a router to end the BGP peering session, a notification message is sent to the BGP neighbor and the connection is closed.

Internal and External BGP

Internal BGP (IBGP) is a BGP peering relationship between routers in the same autonomous system. External BGP (EBGP) is a BGP peering relationship between routers in different autonomous systems. BGP treats updates from internal peers differently than updates from external peers.

Before any BGP speaker can peer with a neighbor router, that neighbor must be statically defined. A TCP session must be established, so the IP address used to peer with must be reachable.

In Figure 6-2, Routers A and B are EBGP peers. Routers B, C, and D are IBGP peers.

Figure 6-2 Identifying EBGP and IBGP Peers

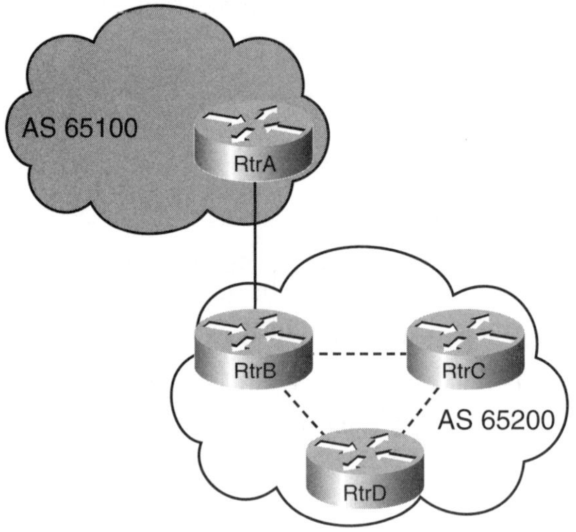

BGP Next-Hop Selection

The next hop for a route received from an EBGP neighbor is the IP address of the neighbor that sent the update.

When a BGP router receives an update from an EBGP neighbor, it must pass that update to its IBGP neighbors without changing the next-hop attribute. The next-hop IP address is the IP address of an edge router belonging to the next-hop autonomous system. Therefore, IBGP routers must have a route to the network connecting their autonomous system to that edge router. For example, in Figure 6-3, RtrA sends an update to RtrB, listing a next hop of 10.2.2.1, its serial interface. When RtrB forwards that update to RtrC, the next-hop IP address will still be 10.2.2.1. RtrC needs to have a route to the 10.2.2.0 network to have a valid next hop.

To change this behavior, use the **neighbor** [*ip address*] **next-hop-self** command in BGP configuration mode. In Figure 6-3, this configuration goes on RtrB. After you give this command, RtrB advertises its IP address to RtrC as the next hop for networks from AS 65100, rather than the address of RtrA. Thus, RtrC does not have to know about the external network between RtrA and RtrB (network 10.2.2.0).

Figure 6-3 BGP Next-Hop Behavior

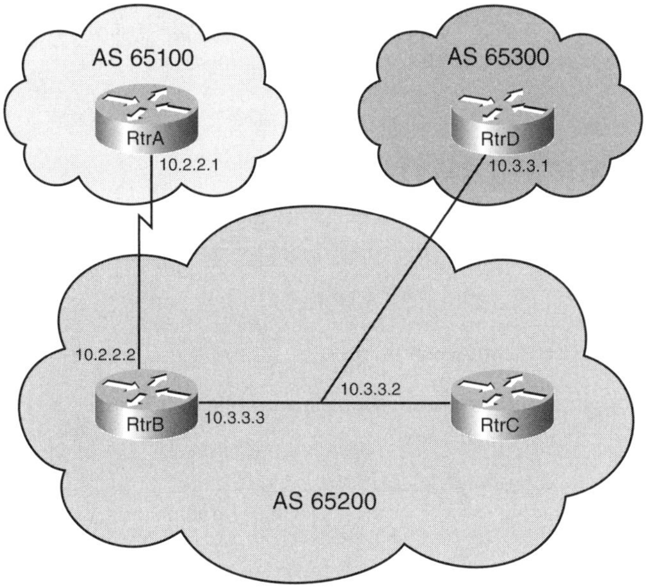

BGP Next Hop on a Multiaccess Network

On a multiaccess network, BGP can adjust the next-hop attribute to avoid an extra hop. In Figure 6-3, RtrC and RtrD are EBGP peers, and RtrC is an IBGP peer with RtrB. When C sends an update to D about network 10.2.2.0, it normally gives its interface IP address as the next hop for D to use. But because B, C, and D are all on the same multiaccess network, it is inefficient for D to send traffic to C, and C to then send it on to B. This process unnecessarily adds an extra hop to the path. So, by default, RtrC advertises a next hop of 10.3.3.3 (RtrB's interface) for the 10.2.2.0 network. This behavior can also be adjusted with the **neighbor** [*ip address*] **next-hop-self** command.

BGP Synchronization Rule

The BGP synchronization rule requires that when a BGP router receives information about a network from an IBGP neighbor, it does not use that information until a matching route is learned via an IGP or static route. It also does not advertise that route to an EBGP neighbor unless a matching route is in the routing table. In Figure 6-3, if RtrB advertises a route to RtrC, then RtrC does not submit it to the routing table or advertise it to RtrD unless it also learns the route from some other IGP source.

Recent IOS versions have synchronization disabled by default. It is usually safe to turn off synchronization when all routers in the autonomous system run BGP. To turn it off in earlier IOS versions, use the command **no synchronization** under BGP router configuration mode.

Configuring BGP

Before beginning to configure BGP, gather the network requirements you need, which should include the following:

- Whether you need to run IBGP for internal connectivity
- External connectivity to the ISP
- Configuration parameters such as neighbor IP addresses and their AS number, and which networks you will advertise via BGP

Table 6-1 lists the basic BGP configuration commands and their functions.

Table 6-1 Basic BGP Configuration Commands

Command	Description
router bgp *AS-number*	Starts the BGP routing process on the router.
neighbor *ip-address* **remote-as** *AS-number*	Sets up peering between BGP routers. IP address must match the source of routing updates.
neighbor peer-*group-name* **peer-group**	Creates a peer group to which you can then assign neighbors.
neighbor *ip-address* **peer-group** *peer-group-name*	Assigns a neighbor to a peer group.
neighbor *ip-address* **next-hop-self**	Configures a router to advertise its connected interface as the next hop for all routes to this neighbor.
neighbor *ip-address* **update-source** *interface-type number*	Configures a router to use the IP address of a specific interface as the source for its advertisements to this neighbor.
no synchronization	Turns off BGP synchronization.
network *prefix* [**mask** *subnet-mask*]	Initiates the advertisement of a network in BGP.

BGP Network Command

In most IGPs, the network command starts the routing process on an interface. In BGP, the command tells the router to originate an advertisement for that network. The network does not have to be connected to the router; it just has to be in the routing table. In theory, it can even be a network in a different autonomous system (not usually recommended).

When advertising a network, BGP assumes you are using the default classful subnet mask. If you want to advertise a subnet, you must use the optional keyword **mask** and specify the subnet mask to use. Note that this is a subnet mask, not the inverse mask used by OSPF and EIGRP network statements. The routing table must contain an exact match (prefix and subnet mask) to the network listed in the network statement before BGP advertises the route.

BGP Peering

BGP assumes that external neighbors are directly connected and that they are peering with the IP address of the directly connected interface of their neighbor. If not, you must tell BGP to look more than one hop away for its neighbor, with the **neighbor** *ip-address* **ebgp-multihop** *number-of-hops* command. You might use this command if you are peering with loopback interface IP addresses, for instance. BGP assumes that internal neighbors might not be directly connected, so this command is not needed with IBGP. If you do peer with loopback IP addresses, you must change the source of the BGP packets to match the loopback address with the **neighbor** *ip-address* **update-source** *interface* command.

To take down the peering session with a neighbor but keep the neighbor configuration, use the **neighbor** *ip-address* **shutdown** command.

BGP Peering States

The command **show ip bgp neighbors** shows a list of peers and the status of their peering session. This status can include the following states:

- **Idle:** No peering; router is looking for neighbor. Idle (admin) means that the neighbor relationship has been administratively shut down.
- **Connect:** TCP handshake completed.
- **OpenSent, or Active:** An open message was sent to try to establish the peering.
- **OpenConfirm:** Router has received a reply to the open message.
- **Established:** Routers have a BGP peering session. This is the desired state.

You can troubleshoot session establishment with debug commands. Use **debug ip bgp events** or **debug ip bgp ipv4 unicast** (in IOS versions 12.4 and up) to see where the process fails. Some common failure causes include AS number misconfiguration, neighbor IP address misconfiguration, a neighbor with no neighbor statement for your router, and a neighbor with no route to the source address of your router's BGP messages.

BGP Path Selection

IGPs, such as EIGRP or OSPF, choose routes based on lowest metric. They attempt to find the shortest, fastest way to get traffic to its destination. BGP, however, has a different way of route selection. It assigns various attributes to each path; these attributes can be administratively manipulated to control the path that is selected. It then examines the value of these attributes in an ordered fashion until it can narrow all the possible routes down to one path.

BGP Attributes

BGP chooses a route to a network based on the attributes of its path. Four categories of attributes exist as follows:

- **Well-known mandatory:** Must be recognized by all BGP routers, present in all BGP updates, and passed on to other BGP routers. For example, AS path, origin, and next hop.
- **Well-known discretionary:** Must be recognized by all BGP routers and passed on to other BGP routers but need not be present in an update, for example, local preference.
- **Optional transitive:** Might or might not be recognized by a BGP router but is passed on to other BGP routers. If not recognized, it is marked as partial, for example, aggregator, community.
- **Optional nontransitive:** Might or might not be recognized by a BGP router and is not passed on to other routers, for example, Multi-Exit Discriminator (MED), originator ID.

Table 6-2 lists common BGP attributes, their meanings, and their category.

Table 6-2 BGP Attributes

Attribute	Meaning
AS path	An ordered list of all the autonomous systems through which this update has passed. Well-known, mandatory.
Origin	How BGP learned of this network. i = by *network* command, e = from EGP, ? = redistributed from other source. Well-known, mandatory.
Local Preference	A value telling IBGP peers which path to select for traffic leaving the AS. Default value is 100. Well-known, discretionary.
Multi-Exit Discriminator (MED)	Suggests to a neighboring autonomous system which of multiple paths to select for traffic bound into your autonomous system. Lowest MED is preferred. Optional, non-transitive.
Weight	Cisco proprietary, to tell a router which of multiple local paths to select for traffic leaving the AS. Highest weight is preferred. Only has local significance.

BGP Path Selection Criteria

BGP tries to narrow its path selection down to one best path; it does not load balance by default. To do so, it examines the path attributes of any loop-free, synchronized (if synchronization is enabled) routes with a reachable next-hop in the following order:

1. Choose the route with the highest weight.
2. If weight is not set, choose the route with the highest local preference.
3. Choose routes that this router originated.
4. Choose the path with the shortest Autonomous System path.
5. Choose the path with the lowest origin code (i is lowest, e is next, ? is last).
6. Choose the route with the lowest MED, if the same Autonomous System advertises the possible routes.
7. Choose an EBGP route over an IBGP route.
8. Choose the route through the nearest IGP neighbor as determined by the lowest IGP metric.
9. Choose the oldest route
10. Choose a path through the neighbor with the lowest router ID.
11. Choose a path through the neighbor with the lowest IP address.

To enable BGP to load balance over more than one path, you must enter the command **maximum-paths** *number-of-paths*. BGP can load balance over a maximum of six paths.

Influencing BGP Path Selection

BGP was not created to be a fast protocol; it was created to enable as much administrative control over route path selection as possible. Path selection is controlled by manipulating BGP attributes, usually using route maps. You can set a default local preference by using the command **bgp default local-preference** and a default MED for redistributed routes with the **default-metric** command under the BGP routing process. But by using route maps, you can change attributes for certain neighbors only or for certain routes only. The earlier section on route maps contains an example of using a route map to set a local preference of 200 for specific redistributed routes. This is higher than the default local preference of 120, so routers within the AS are more likely to prefer that path than others.

Route maps can also be applied to routes sent to or received from a neighbor. The following example shows a simple route map that sets a MED value and adds two more copies of its AS number to the AS path on all routes advertised out to an EBGP neighbor:

```
route-map MED permit 10
 set metric 50
 set as-path prepend 65001 65001
!
router bgp 65001
 neighbor 10.1.1.1 route-map MED out
```

When attributes are changed, you must tell BGP to apply the changes. Either clear the BGP session (**clear ip bgp** *) or do a soft reset (**clear ip bgp * soft in | out**). Routers using recent IOS versions do a route refresh when the session in cleared inbound.

Filtering BGP Routes

You can combine route maps with prefix lists to filter the routes advertised to or received from a BGP peer, to control routes redistributed into BGP, and to set BGP attributes for specific routes. Prefix lists alone can be applied to a neighbor to filter route updates.

To use a prefix list, plan the implementation by determining the requirements. Then create a prefix list to match the networks to be filtered. Permit the networks you want to allow to be advertised and deny all others. Next apply the prefix list to the BGP neighbor, inbound or outbound. The next example shows a prefix list that permits only summary routes in the 172.31.0.0 network. All other routes are denied by default. The prefix list is then applied to BGP neighbor 10.1.1.1 outbound, so only these routes will be advertised to that peer:

```
ip prefix-list Summary permit 172.31.0.0/16 le 20
!
router bgp 65001
neighbor 10.1.1.1 prefix-list Summary out
```

To verify the results of your configuration use the command **show ip prefix-list**. To clear the counters shown in that command, use the **clear ip prefix-list** command.

Combine a prefix list with a route map to set attributes on the routes allowed in the prefix list. In the following example, prefix list Summary is used again. A route map sets the Med for those routes to 100 when they are advertised. It sets a Med of 200 for all other routes advertised. The route map is then applied to BGP neighbor 10.1.1.1 outbound:

```
route-map CCNP permit 10
match ip address prefix-list Summary
set metric 100
route-map CCNP permit 20
set metric 200
!
router bgp 65001
neighbor 10.1.1.1 route-map CCNP out
```

BGP Authentication

BGP supports MD5 authentication between neighbors, using a shared password. It is configured under BGP router configuration mode with the command **neighbor** {*ip-address* | *peer-group-name*} **password** *password.* When authentication is configured, BGP authenticates every TCP segment from its peer and checks the source of each routing update. Most ISPs require authentication for their EBGP peers.

Peering succeeds only if both routers are configured for authentication and have the same password. If a router has a password configured for a neighbor, but the neighbor router does not, a message such as the following displays on the console while the routers attempt to establish a BGP session between them:

```
%TCP-6-BADAUTH: No MD5 digest from [peer's IP address]:11003 to
  [local router's IP address]:179
```

Similarly, if the two routers have different passwords configured, a message such as the following will display on the screen:

```
%TCP-6-BADAUTH: Invalid MD5 digest from [peer's IP
  address]:11004 to [local router's IP address]:179
```

Verifying BGP

One of the best commands to verify and troubleshoot your BGP configuration is **show ip bgp** to see the BGP topology database. This is such an important command that it's worth looking at in depth. The command output lists a table of all the networks BGP knows about, the next hop for each network, some of the attributes for each route, and the AS path for each route. The sample output from this command was taken from an actual Internet BGP peer.

```
route-server>show ip bgp
BGP table version is 22285573, local router ID is 12.0.1.28
Status codes: s suppressed, d damped, h history, * valid, > best, i
  - internal,
       r RIB-failure, S Stale
Origin codes: i - IGP, e - EGP, ? - incomplete

   Network        Next Hop           Metric LocPrf Weight Path
* 3.0.0.0        12.123.137.124                       0     7018 2914 9304
  80 i
*>               12.123.1.236                         0     7018 2914 9304
  80 i
* 3.51.92.0/23 12.123.137.124                       0     7018 ?
*                12.122.125.4       2366              0     7018 ?
*>               12.123.1.236                         0     7018 ?
* 8.6.6.0/24     12.123.137.124                       0     7018 701 14744
  14744 14276 i
*                12.123.145.124                       0     7018 701 14744
  14744 14276 i
*>               12.123.1.236                         0     7018 701 14744
  14744 14276 i
```

Networks are listed in numerical order, smallest to largest. The first three columns list each route's status. An asterisk (*) in the first column means that the route has a valid next hop. Some other options for the first column include the following:

- **"s" for suppressed:** BGP knows about this network but is not advertising it, usually because it is part of a summarized route.
- **"d" for dampened:** BGP can stop advertising a network that flaps (goes up and down) too often until it is stable for a period of time.
- **"h" for history:** BGP knows about this network but does not currently have a valid route to it.
- **"r" for RIB failure:** The route was advertised to BGP but it was not installed in the IP routing table. This might be because of another protocol having the same route with a better administrative distance.
- **"S" for stale:** Used with nonstop forwarding to indicate that the route is stale and needs to be refreshed when the peer is reestablished.

The second column has a greater-than sign (>) beside the route that was selected as the best path to that network. In the example, the second route was selected for network 3.0.0.0.

The third column is blank in the example, which means that the router learned all the routes from an external neighbor. A route learned from an IBGP neighbor would have an "I" in the third column.

The fourth column lists the networks. Those without a subnet mask, such as network 3.0.0.0, use their classful mask. As seen in the example, when the router learns about the same network from multiple sources, it lists only the network once.

The fifth column lists the next-hop address for each route. As you learned in the previous sections on BGP next hops, this might or might not be a directly connected router. A next-hop of 0.0.0.0 means that the local router originated the route.

If a Med value was received with the route, it is listed in the Metric column. Notice that the advertisement for network 3.51.92.0/23 from the router at 12.122.125.4 has a large Med value of 2366. Because the default Local Preference is used for each of the routes shown, no local preference value is displayed. The default Weight value of 0 is listed, however.

The ninth column shows the AS path for each network. Reading this field from left to right, the first AS number shown is the adjacent AS this router learned the route from. After that, the AS paths that this route traversed are shown in order. The last AS number listed is the originating AS. In the example, our router received an advertisement about network 3.0.0.0 from its neighbor AS 7018, which heard about it from AS 2914, which heard about it from AS 9304. And AS 9304 learned the route from AS 80, which originated it. A blank AS path means that the route was originated in the local AS.

Note

In the AS Path column, note that network 8.6.6.0 shows AS 14744 twice in its AS path list. Most likely AS 14744 has prepended an extra copy of its AS number to make the path through it less attractive than the path through other autonomous systems. In this case it did not work because the only paths to 8.6.6.0 this router knows about all go through AS 14744.

The last column shows how BGP originally learned about the route. Networks 3.0.0.0 and 8.6.6.0 show an "i" for their origin codes. This means that the originating router had a network statement for that route. Network 3.51.92.0 shows a "?" as its origin. This means that the route was redistributed into BGP; BGP considers it an "incomplete" route. You will likely never see the third possibility, an "e," because that means BGP learned the route from the Exterior Gateway Protocol (EGP), which is no longer in use.

Some other useful commands for verifying and troubleshooting BGP include

- **show ip bgp rib-failure:** Displays routes that were not inserted into the IP routing table and the reason they were not used.
- **show ip bgp summary:** Displays the memory used by the various BGP databases, BGP activity statistics and a list of BGP neighbors.
- **show ip bgp neighbors:** Displays details about each neighbor. Can be modified by adding the neighbor IP address.
- **show ip bgp neighbors** *address* **[received | routes | advertised]:** Lets you monitor the routes received from and advertised to a particular neighbor.

You can search for "Internet route servers" to find listings of BGP routers that enable public telnet access for viewing their BGP tables. Trying some of these commands on a public route server can help you become familiar with them.

CHAPTER 7

Branch Office Connectivity

The needs of branch offices are changing. This is due to the adoption of unified networks that support voice, video, and data; the consolidation of IT resources; and the physical mobility of many users.

Branch Office Design Considerations

Some design considerations for branch offices include

- **Connectivity technologies:** What WAN options are available?
- **Resiliency:** How much downtime can the site tolerate? Are there alternate WAN paths available?
- **Routing:** Will the WAN support routing protocols?
- **Services:** Are services such as NAT, WAN optimization, and QoS needed at the branch?
- **Security and compliance:** What security is needed, where will it be placed, and how will that affect routing?
- **Mobility:** Do teleworkers use this branch for VPN access?

Strive for a consistent design across your branch offices, with a structured method of handling change management and configuration. Branch offices have different needs from campus locations, but you can still have a common design foundation by creating standard designs for different size offices. Each category of branch office is its own "place in the network." Categorize offices not only by the number of users they have, but also by how critical the branch is. The following office profiles are meant as a baseline, not a recommendation for every network.

Small Branch Office Design

A small branch office typically leverages an ISR router to provide multiple services such as WAN and PSTN connectivity, NAT, WAN optimization, firewall, and DHCP. Its WAN connectivity might be a T1 primary link with a cable or DSL backup link using an IPsec VPN. You might run a routing protocol or simply use floating static routes. The infrastructure typically consists of Layer 2 switching—either internal to the router or using an external switch, computers, phones, and printers.

This design is cost-effective and provides minimum devices to manage. However, network resiliency suffers because the router is a single point of failure.

Medium Branch Office Design

A medium-sized branch office requires some additional resiliency and network equipment. There typically are redundant WAN routers with dual connections to a private WAN using either MPLS or Frame Relay. The routers will be higher capacity devices but might still provide services such as firewall, NAT, DHCP, and WAN optimization. The network might use a FHP such as HSRP. The infrastructure typically consists of either Layer 2 or Layer 3 external switches, computers, phones, and printers.

This design is more resilient than the small office design. However, the dual paths add path control complexity and dynamic routing is needed to accomplish load sharing across the links. Documenting traffic flows becomes more important.

Large Branch Office Design

A large branch office is similar to a campus design in that it typically uses a layered design with redundancy at all but the access layer. Stand-alone devices for firewalls and WAN optimization might be used, along with multilayer switches. This branch can provide services to other branches and can thus benefit from an MPLS WAN with its any-to-any connectivity. The infrastructure is engineered for high availability. It typically consists of dual WAN access routers, dual distribution switches, and dual firewalls.

This design adds more complexity to the routing structure and might require route redistribution and filtering. Regulatory requirements can lead to deploying overlay VPNs such as Dynamic Multipoint VPN (DMVPN) or Group Encrypted Transport VPN (GET VPN).

Implementing Branch Offices

A full branch office implementation plan aims to integrate the new design without disrupting the existing users. It would include the following items:

- Deployment strategy
- Network diagrams
- Installation and site tests
- Site survey results
- Installation guidelines
- Device-specific configuration templates
- Test and acceptance plan
- Documentation of the new network

Verifying Existing Services

Because a CCNP-level engineer should handle the device configuration, we start at that step of the plan. The implementation is likely an upgrade to an existing branch office network, so the first step is to identify and document the current configuration of every device and the services it provides. These services might include NAT, HSRP, ACLs, firewall, or redirection services such as PBR or WCCP. The IP address schema must also be documented.

To see the NAT settings, you might use the commands **show ip nat translations** and **show ip nat statistics**. To document DHCP, use the commands **show ip dhcp pool** and **show ip dhcp server statistics**. The **show access-lists** and **show ip interface** commands give you information about ACLs and where they are applied. To see what IOS firewall rules are in place, use the commands **show ip inspect interfaces** and **show zone-pair security**. Verify HSRP operation with **show standby brief**. The command **show ip policy** displays any PBR policies.

Configuring a Backup DSL Connection

Assume that this branch already has a WAN connection, and you are adding redundancy by provisioning a backup DSL connection. Voice does not use all the available bandwidth on a phone line; it uses frequencies up to only approximately 3 kHz. DSL was created to use the space between 3 kHz and 1 MHz to send data traffic over a telephone local loop. Thus, both voice and data can be sent simultaneously over the same connection. (Some variants of DSL use the entire spectrum, however, so no voice can be sent.) DSL is a physical layer medium that extends between the subscriber's DSL modem and the provider's DSL Access Multiplexer (DSLAM).

Asymmetrical DSL has higher downstream (from the provider's Central Office to the subscriber) bandwidth than upstream (from the subscriber to the CO.) Symmetrical DSL has the same bandwidth both downstream and upstream. You sometimes see these referred to as "asynchronous" and "synchronous" DSL.

The various types of DSL include

- **ADSL:** Asymmetric DSL supports both voice and data. Downstream bandwidth goes up to 8 Mbps; upstream goes up to 1 Mbps. Two other versions, ADSL2 and ADSL2+, provide 24 Mbps downstream and 1.5 Mbps upstream. The maximum distance from the CO is 18,000 feet, or 5.46 km.
- **VDSL:** Very-high-rate DSL can be either symmetric or asymmetric and can carry voice along with data. Maximum symmetric bandwidth is 26 Mbps; maximum asymmetric is 52 Mbps downstream and 13 Mbps upstream. The maximum distance from the CO is 4,500 feet, or 1.37 km.
- **SDSL:** Symmetric DSL carries only data, with a maximum for both downstream and upstream of 768 kbps. The distance limitation is 22,000 feet, or 6.7 km. It is a proprietary technology that uses only one twisted pair of wires.
- **HDSL:** High-data-rate DSL uses two twisted pairs of wires to achieve a maximum symmetrical bandwidth of 2.048 Mbps. Its maximum distance from the CO is 12,000 feet, or 3.7 km. HDSL carries only data, no voice.
- **G.SHDSL:** Symmetric High-speed DSL has a symmetrical data rate of 2.3 Mbps and the longest maximum distance: 28,000 feet, or 8.52 km. It also carries only data, no voice.

Figure 7-1 shows how ADSL components work together in a typical residential implementation. The telephone company's Central Office forwards both POTS and DSL data traffic over the same line to the subscriber. The line enters at the Network Interface Device (NIDS) and branches toward the telephone and the PC. A low-pass filter blocks everything but voice frequencies from reaching the phone. A DSL modem (or router with a DSL interface) forwards data to the PC. When the Central Office receives traffic from the subscriber, a splitter sends voice frequencies to the PSTN switch and DSL frequencies to the DSLAM. The DSLAM sends data traffic to a router for forwarding to the Internet.

Figure 7-1 Components of an ADSL System

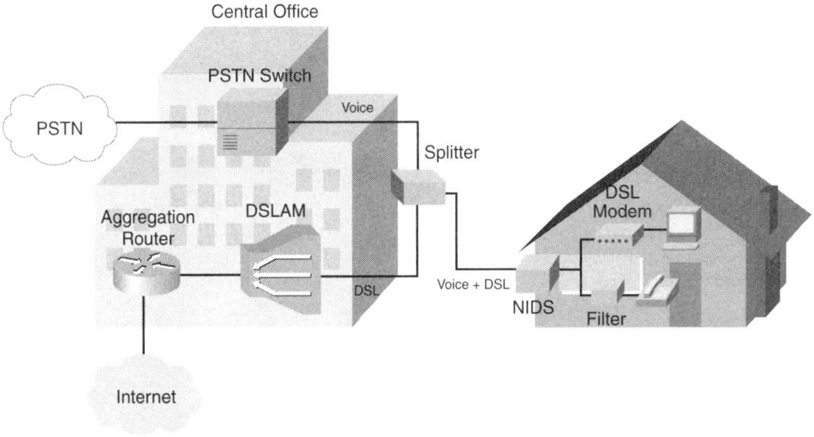

Recall that DSL is a Layer 1—Physical Layer—technology. Following are the three methods of carrying data at Layer 2 over DSL:

- **Bridging:** Based on RFCs 1483 and 2684. Ethernet traffic is just bridged from the subscriber PCs, through the DSL modem and the DSLAM, to a provider router. Is not as secure or scalable as other methods.
- **Point-to-Point Protocol over Ethernet (PPPoE):** The most common Layer 2 method of carrying data over DSL. PPP traffic is encapsulated in Ethernet frames.
- **Point-to-Point Protocol over ATM (PPPoA):** PPP packets are routed over ATM between the subscriber equipment and the provider.

In the following example we use PPoA, which requires a CPE router because traffic is routed from the subscriber PCs to the aggregation router. The PPP session is established between the CPE router and the aggregation router. Either Password Authentication Protocol (PAP) or Challenge Handshake Authentication Protocol (CHAP) authentication can be used. Multiple users are supported if the CPE router is configured to do DHCP and NAT. Traffic between the CPE router and the aggregation router is encapsulated as ATM at Layer 2.

When configuring PPPoA you must set up the internal Ethernet interface, a dialer interface, NAT or PAT, DHCP, and a static default route. Because this is ATM, you must configure Virtual Path Identifier (VPI) and Virtual Circuit Identifier (VCI) information on the external interface to match that of the provider. The type of ATM encapsulation must be specified, PPPoA must be enabled, and the ATM interface must be linked to the virtual dialer interface. A dialer pool is associated with PVC. The final configuration is shown in the following examples.

Internal Ethernet interface:

```
interface FastEthernet0/1
 description Internal interface
 ip address 172.16.1.1 255.255.255.0
 ip nat inside
```

Dialer interface:

```
interface Dialer1
 ip address negotiated
 ip mtu 1492
 ip nat outside
 encapsulation ppp
 dialer pool 1
 ppp authentication chap
 ppp chap password 0 dslpass
```

Port address translation (PAT)

```
access-list 100 permit ip 172.16.1.0 0.0.0.255 any
ip nat inside source list 100 interface Dialer1 overload
```

DHCP:

```
ip dhcp pool Users
  import all
  network 172.16.1.0 255.255.255.0
  default-router 172.16.1.1
```

Static default route:

```
ip route 0.0.0.0 0.0.0.0 Dialer1
```

External ATM interface:

```
interface ATM1/0
 description DSL interface
 no ip address
 dsl operating-mode auto
 pvc 1/100
 encapsulation aal5mux ppp dialer
 dialer pool-member 1
```

To verify and troubleshoot the DSL configuration, use the commands **show dsl interface** atm *number*, **debug atm events**, **debug ppp authentication**, **show ip route**, **ping**, and **traceroute**.

Configuring an IPsec VPN

IPsec is not covered in depth on the ROUTE exam, but you need to understand it well enough to verify the configuration and add routing across it. This sample branch uses an IPsec VPN to connect to the headquarters when the backup DSL link is active.

When IPsec establishes a VPN between two peer hosts, it sets up a security association (SA) between them. SAs are unidirectional, so each bidirectional data session requires two. The Internet Security Association and Key Management Protocol (ISAKMP) defines how SAs are created and deleted.

An IPsec transform set defines how VPN data will be protected by specifying the IPsec protocols that will be used. You can specify up to four transforms, and the algorithm to use with each. You can also configure either tunnel or transport mode. (Tunnel is default.)

You use a crypto ACL to identify traffic that should be protected by the IPsec VPN. Any traffic permitted in the ACL will be sent over the VPN. Traffic denied by the ACL will not be dropped; it will simply be sent normally.

A crypto map pulls together the transform sets and crypto ACLs and associates them with a remote peer. After the crypto map is configured, it must be applied to an interface for it to take effect. It is applied at the *outgoing* interface—the one that VPN traffic uses to reach the other end of the VPN. You might need to use a static route or otherwise adjust your routing to force traffic bound for the VPN destination networks to use the correct outgoing interface.

To verify the IPSEC VPN, use the following commands:

- **show crypto map:** Shows the crypto ACLs, any peers, and the interface where the crypto map is applied
- **show crypto isakmp sa:** Shows information about the ISAKMP security associate negotiation process
- **show crypto ipsec sa:** Shows the settings used by current SAs, including tunnel status and peers

Configuring a Floating Static Route

The IPsec tunnel can be used solely as a backup link, or you can load balance between it and the primary link. To use it as a backup link, you can configure a floating static route. A floating static route is one with an administrative distance greater than the primary route. If the primary route is active, the static route will not be placed into the routing table due to its higher AD. But when the primary route is down, the static route will be used.

The command syntax for a floating static route is: **ip route** *destination-network next-hop-address administrative-distance*. You can find the AD of the primary route by using the command **show ip protocols**. EIGRP has an AD of 90, so you might use 100 as the AD of a floating static route when the primary route is learned via EIGRP.

Configuring Dynamic Routing over a GRE Tunnel

To use the IPsec tunnel as an "always on" connection, you need to send routing updates over it. However, IPsec VPNs do not carry broadcast or multicast traffic. You need to create a tunnel within the IPsec tunnel to carry the routing traffic. Four ways to do this include

- **DMVPN:** Creates multipoint tunnels on-demand. Good for scenarios when spoke-to-spoke connections are needed.
- **GET VPN:** Creates encrypted multipoint tunnels on-demand. Good for scenarios when secure spoke-to-spoke connections are needed.
- **Virtual Tunnel Interface (VTI):** Creates an always-on tunnel that carries unicast and multicast traffic. Enables you to configure the routing protocol on the tunnel interface, saving the 4 extra bytes required for a GRE header.
- **Generic Routing Encapsulation (GRE):** GRE is a tunneling protocol that can support multiple Layer 3 protocols. It enables the use of multicast routing protocols across the tunnel. It adds a 20-byte IP header and a 4-byte GRE header. GRE does not encrypt traffic or use any strong security measures to protect the traffic. GRE can be used along with IPsec to provide data source authentication, data confidentiality, and assurance of data integrity. GRE over IPsec tunnels are typically configured in a hub-and-spoke topology over an untrusted WAN to minimize the number of tunnels that each router must maintain.

In this section, we configure a GRE tunnel to carry EIGRP traffic over the IPSEC tunnel. This basically creates a tunnel within a tunnel, as shown in Figure 7-2.

Figure 7-2 GRE over IPSEC Tunnel

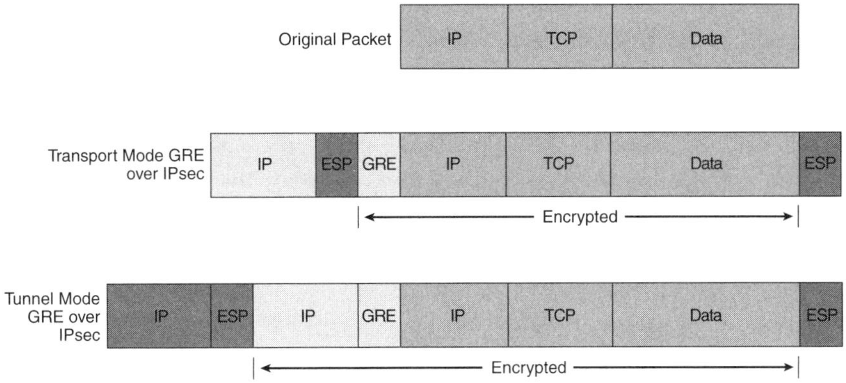

Configuring a GRE tunnel is fairly easy. Follow these steps on the routers at each end of the tunnel:

1. Create a loopback interface to use as the tunnel endpoint. Using a loopback rather than a physical interface adds stability to the configuration.
2. Create the GRE tunnel interfaces.
3. Add the tunnel subnet to the routing process so that it exchanges routing updates across that interface.
4. Add GRE traffic to the crypto access list, so that IPsec encrypts the GRE tunnel traffic.

The following example shows a tunnel interface configured for GRE. The **mode** command is shown only as a reference; because it is the default, it would not normally appear in the configuration:

```
interface Tunnel1
 ip address 172.16.5.2 255.255.255.0
 tunnel source Serial0/0
 tunnel destination 10.1.1.1
 tunnel mode gre ip
```

To verify the configuration, use the **show ip route** command and look for the remote routes in the routing table. They should have a next hop of the tunnel interface. A **traceroute** shows traffic going across the tunnel. The

show crypto ipsec sa command output shows increased traffic as the traceroute goes through the tunnel.

Load Sharing with EIGRP

In the EIGRP chapter you saw how EIGRP can load balance across unequal-cost links. Because you have an always-on IPsec tunnel, you might want to use that in addition to your primary route. Use the **variance** *multiplier* command under EIGRP configuration mode. Look in the EIGRP topology table to find the metric for the best route and the secondary route. Determine what you need to multiply the best route's metric by for it to be more than the secondary route's metric. That is the variance multiplier number that you should use to configure EIGRP to load balance over both paths.

CHAPTER 8

Mobile Worker Connectivity

Mobile workers might be in a home office, with an always-on secure connection to the corporate network, managed centrally by the corporation (business-ready mobile worker.) Or they might be truly mobile, connecting via a laptop or public computer to the corporate network (traditional mobile worker.) In planning for either type of mobile worker, consider the network access technologies and infrastructure services needed, such as the following:

- **Bandwidth requirements:** Because mobile workers use the same applications as office workers—email, other applications, voice, video, real-time collaboration—they need sufficient bandwidth. Typical remote access technologies include residential cable, DSL, and wireless.
- **Connection security:** Use site-to-site VPNs for permanent home users and remote access VPNs for mobile users. These can be either IPsec or Secure Sockets Layer (SSL) VPNs.
- **Corporate security:** Because a remote environment is less controlled than an office environment, use firewalls, intrusion prevention services (IPS), and URL filtering to protect the corporate network from remote users.
- **User authentication:** Use network access control (NAC), AAA servers, or other authentication mechanisms to protect access to corporate resources.
- **QoS:** If voice and video are used, determine how you will prioritize that traffic, and how you will address the differences in upload and download speeds of common broadband connections.
- **Management:** Support for remote workers is more complex when they are not under corporate control. Provide methods to push security policies and updates to mobile workers.

Table 8-1 compares the two types of mobile workers and the ability to offer these services to each.

Table 8-1 Comparison of Traditional and Business-Ready Mobile Workers

	Traditional Mobile Worker	Business-Ready Mobile Worker
Access to applications and services	Basic	Full
Voice and video support	Limited to none	Yes
QoS	No (best effort)	Yes
Security	Relies on end-user	Controlled by corporate IT
Remote management	No	Yes

Components of a Mobile Worker Solution

There are three major groups of components in a mobile worker solution: corporate devices, devices located at the remote site, and optional additional services.

The corporate components include headend routers, devices to terminate the VPNs such as ASA firewalls, authentication services, and central management devices.

The remote site components for a Business-ready solution include broadband access, a VPN router with QoS capabilities, and a computer. There can also be a wireless access point, an IP phone, and a video telephony camera. A traditional mobile worker might have only a laptop, perhaps with a softphone on it.

The corporation might offer additional services to the mobile worker, such as IP telephony, voice mail, or contact center.

The corporation also needs to decide which type of VPN services to offer: IPsec, SSL, or both. IPsec requires a client on the endpoint (computer or router), but it is a well-proven technology that provides full access to all network applications. It is a good choice for mobile workers using company-managed devices.

SSL is also a well-proven technology and can be used with or without a client. When used from a web browser, it provides access even from nonmanaged devices, and the portals can be customized to provide appropriate access for employees or business partners. Some applications might not work well through the web portal, however.

Implementing a Mobile Worker Solution

Some of the things you need to implement in a mobile worker solution are

- The VPN solution, either IPsec or SSL
- A firewall solution, for stateful Layer 3–7 protection
- Intrusion prevention to defend against worms and viruses
- Wireless security, if wireless is used, with encryption and 802.1x authentication.
- QoS for voice and video
- Ports at the remote site for printers and other devices
- PoE ports for IP phones

Cisco Easy VPN can simplify the deployment of all these services and policies to remote users. Easy VPN enables a server to push down VPN configuration to a client. It is a way to create site-to-site VPNs without manually configuring each remote router. Thus it is good for remote sites without technical support. It can also be used with software clients for remote users.

Cisco Easy VPN dynamically handles the following items:

- Negotiating VPN tunnel parameters
- Establishing the VPN tunnel based on those parameters
- NAT, PAT, or ACL configuration
- User authentication
- Managing encryption and decryption keys
- Authenticating, encrypting, and decrypting traffic

Cisco Easy VPN has two components: a server and a remote client. The *Easy VPN Server* can be a Cisco router, ASA Firewall, or Cisco VPN concentrator. It contains security policies and pushes those to remote clients. The *Easy VPN Remote* can be a Cisco router, ASA Firewall, a hardware client, or a software client. It contacts the server and receives policies from it to establish the IPsec tunnel. The steps to configure the headend for Easy VPN are

1. Allow IPsec traffic through the edge firewall or access list.
2. Create the IP address pool for the VPN clients.
3. Verify the IPsec VPN configuration.

4. Ensure that corporate devices have routes to the VPN subnets.
5. Tune NAT to bypass VPN traffic.

Allow IPsec Traffic

IPsec uses the following ports and protocols:

- Encapsulating Security Payload (ESP)—IP protocol 50
- Authentication Header—IP protocol 51
- ISAKMP—UDP port 500
- NAT Traversal (NAT-T)—UDP port 4500

These protocols and ports must be allowed in through the firewall from the outside for an IPsec tunnel to be established. If your network is protected by a firewall appliance or module, coordinate these changes with your security personnel. If your network is protected by an IOS firewall, determine whether you have a zone-based firewall or a classic IOS firewall. The command **show zone-pair security** produces output if you have a zone-based firewall. The command **show ip inspect interfaces** produces output if you have a classic IOS Firewall. The command **show access-lists** will show you any access lists associated with the firewall that will need to be modified to allow IPsec traffic.

Fortunately, configuring IPsec and firewalls is outside the scope of this exam, so you need to understand it only in concept.

Create the Address Pool

VPN clients have a public Internet address until they reach your network but need a private inside address to access network resources. The addresses can be given by a router acting as a DHCP server, by other DHCP servers, or by AAA servers.

Plan your address pool so that you have enough client addresses for the maximum number of users and so that the subnets can be summarized when their routes are advertised to internal devices. Be sure that the client address range is not used anyplace else in the network.

To configure a DHCP pool on a router, use the command:

```
ip local pool pool-name first-address last-address [mask subnet-
  mask]
```

For the exam, you are not required to add this IP address pool to the VPN configuration. In theory you would request a change ticket for your VPN support team to make that change.

Verify the IPsec VPN

You are likewise required to understand only the components of a VPN, not to know how to configure it. The two main things you need to know are the function of a crypto map and the commands to verify IPsec connectivity.

A crypto map is described in Chapter 7, "Branch Office Connectivity." It basically groups the VPN settings to apply them to an interface. It consists of the crypto ACL, which defines the traffic to be encrypted, security policies such as how to protect the data, and other security parameters such as peer IP address or tunnel lifetime. The crypto map is applied to the exit interface for traffic needing to be protected.

Commands to verify IPsec connectivity include

- **show crypto map:** Shows the crypto ACLs, any peers, and the interface where the crypto map is applied
- **show crypto isakmp sa:** Shows information about the ISAKMP security associate negotiation process
- **show crypto ipsec sa:** Shows the settings used by current SAs, including tunnel status and peers
- **show crypto engine connections active:** Shows the status of any IPsec tunnels

Route to the VPN Subnets

Internal resources need to have a route to the VPN subnets. One way to do this is to make the VPN subnet a part of an existing subnet applied to a router interface. This subnet should already be advertised to the rest of the network, so no changes to routing are needed. The disadvantage is that you need to enable proxy-ARP on the router interface connected to the subnet.

A second option is to use a separate subnet for the VPN clients and then inject that subnet into the routing protocol. An IPsec feature called Reverse Route Injection (RRI) does that dynamically. It injects a host route into the routing table for each client while they are connected and then withdraws that route when they disconnect. You would then redistribute those routes into the routing protocol. Alternatively, you can create a static route for the VPN subnets and redistribute the static route into your routing protocol.

Tune NAT

VPN traffic should bypass the NAT process as it leaves your network. If the same edge router is doing NAT and terminating the VPNs, the router processes the NAT service before the IPsec service. It is simple to configure NAT bypass because NAT uses an access list to identify which traffic to translate. Modify the access list to deny traffic destined to the VPN subnet. Be sure to permit all other traffic. You can then match that access list in a route-map and use the route-map to modify the NAT traffic. Your final configuration might look something like this:

```
access-list 101 deny ip any vpn-subnet vpn-mask
access-list 101 permit ip any any
!
route-map NAT permit 10
match ip address 101
!
ip nat inside source route-map NAT pool NAT-POOL overload
```

Be sure to verify your configuration by checking the IP address of a VPN client, pinging internal resources, checking the routing table, and the IPsec SAs.

CHAPTER 9

IPv6 Introduction

IPv6 is an extension of IP with several advanced features:

- Larger address space.
- No more need for NAT.
- Simpler header for increased router efficiency.
- No more broadcasts.
- Stateless autoconfiguration.
- Built-in support for Mobile IP.
- Built-in support for IPsec security.
- Rich transition features.
- Easy IP address renumbering.
- Capability to have multiple addresses per interface.
- Routers create link-local addresses for use by IGPs.
- As with IPv4, the addresses can be provided by the ISP or can be provider independent.

The primary adoption of IPv6 is driven by the need for more addresses. Given the growth in Internet use and the emergence of large groups of Internet users worldwide, this is a significant requirement. Another reason to use IPv6 is growth in the size of the current Internet routing table. IPv4 addresses are not summarized enough to keep the size down, increasing the load on Internet routers. Additionally, although the use of NAT has postponed the need for IPv6, it breaks TCP/IP's end-to-end networking model.

IPv6 Addressing

IPv4 addresses are 32-bits long and written in dotted decimal, whereas IPv6 addresses are 128 bits and written in hexidecimal. They are typically divided into a 64-bit network portion and a 64-bit host portion. The first 48 bits of the network portion are considered as Global Address Space. These bits consist of the following elements (see Figure 9-1):

- The first three bits (/3) of a unicast address are always 001.
- The next 13 bits (/16) identify the Top-Level Aggregator (TLA); the upstream ISP.
- The next 24 bits (/40) identify the next-level aggregator, or regional ISP.

Enterprises are assigned /48 addresses and have 16 bits of subnetting available.

The host portion of the address is last 64 bits. The subnet mask is specified using Classless Interdomain Routing (CIDR) notation. Figure 9-1 shows the address components.

Figure 9-1 IPv6 Address Structure

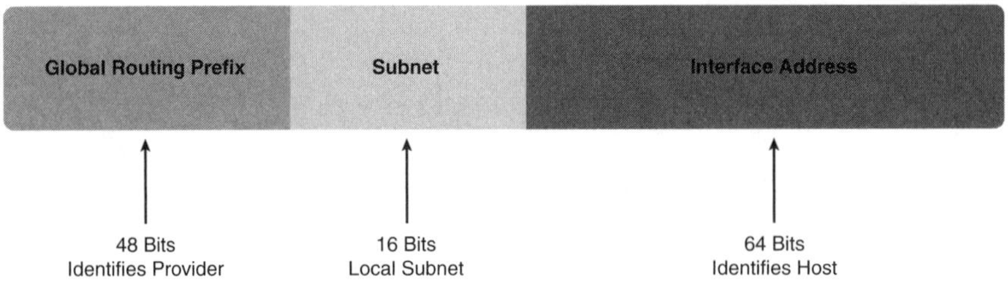

Simplifying an IPv6 Address

There are two ways to shorten the representation of an IPv6 address. Take the example address 2001:0000:0001:0002:0000:0000:0000:ABCD.

- Leading zeros can be omitted. Doing this would shorten the preceding address to 2001:0:1:2:0:0:0:ABCD.
- Sequential zeros can be shown as double colons. *This is allowed only once per address.* Adding this would simplify the above address even further, to 2001:0:1:2::ABCD.

For the exam, be sure that you can distinguish between correct and incorrect IPv6 addresses. For instance, the address 2001::1:2::ABCD is incorrect because it uses double colons twice.

Special Addresses

IPv6 does not support broadcasts but replaces broadcasts with multicasts. IPv6 also uses Anycast, which involves using the same address on two devices. Anycast can be used to implement redundancy and has been back-ported to IPv4.

Each IPv6 system must recognize the following addresses:

- Its unicast addresses
- Link local address (begins with FE80/10)
- Loopback (::1/128)
- All-nodes multicast (FF00::1)
- Site-local multicast (FF02::2)
- Solicited-nodes multicast (FF02::1:FF00/104)
- Default route (::/0)

Additionally, some systems also use the following addresses:

- IPv4 compatible address (::/96 | 32-bit, IPv4 address).
- Second unicast address shared with another system (anycast).
- Additional multicast groups.
- Routers must support subnet-router anycast (all zeros EUI-64).
- Routers must support local all-routers multicast (FF01::2), link-local (FF02::2), and site-local (FF05:2).
- Routers must support routing protocol multicast groups.

IPv6 Host Addressing

An IPv6 host can obtain an IP address by manual assignment, by manually assigning the network address only, by using stateless autoconfiguration, or by using DHCPv6. IPv6 is not enabled by default on Cisco routers. To enable IPv6 routing, the command is **ipv6 unicast-routing** at the global configuration mode.

To ping any IPv6 address, including link-local addresses, use the command **ping ipv6** *destination-address* **source** *exit-interface*. Note that you must specify a source.

Manual IP Address Assignment

To manually assign an IPv6 address to a router interface, use the command **ipv6 address** *ipv6-address/prefix-length*. The following example shows a router interface with two IPv6 addresses. In the first address, note leading zeros are omitted in two of the quartets. In the second address, note the use of the double colons:

```
RouterA# configure terminal
RouterA(config)# ipv6 unicast-routing
!
RouterA(config)# interface fastethernet0/0
RouterA(config-if)# ipv6 address
 2001:0:aabb:1:2222:3333:4444:5555/64
RouterA(config-if)# ipv6 address 2001:0:aabb:2::1 /64
```

Manual Network Assignment

The router can create its own IPv6 address when it knows its network. If the end system has a 64-bit MAC address, it concatenates the network prefix and its MAC address to form an IPv6 address. If the end system has a 48-bit MAC address, it flips the global/local bit (the 7th bit) and inserts 0xFFEE into the middle of the MAC address. The resulting 64-bit number is called the EUI-64 address. The prefix and EUI-64 address are concatenated to form the host IPv6 address. The command is **ipv6 address** *ipv6-prefix::/prefix-length* **eui-64**.

The following example shows this command and the resulting link-local and global unicast address. Note the interface MAC address and how it relates to the IPv6 addresses.

```
RouterA(config)# interface fastethernet0/0
RouterA(config-if)# ipv6 address 2001:8:1234:aabb::/64
!
R1# show int fa 0/0
FastEthernet0/0 is up, line protocol is up
 Hardware is MV96340 Ethernet, address is 001d.a188.33c1 (bia
 001d.a188.33c1)
!
R1# show ipv6 int fa0/0
FastEthernet0/0 is up, line protocol is up
 IPv6 is enabled, link-local address is FE80::21D:A1FF:FE88:33C1
 [TEN]
 No Virtual link-local address(es):
 Global unicast address(es):
  2001:8:1234:AABB:21D:A1FF:FE88:33C1, subnet is
  2001:8:1234:AABB::/64 [EUI/TEN]
```

```
Joined group address(es):
 FF02::1
 FF02::2
```

Stateless Autoconfiguration

One big benefit of IPv6 is stateless autoconfiguration, the capability of a host to automatically acquire an IP address without needing DHCP. It uses its link-local address and the Neighbor Discovery Protocol (NDP) to do this.

Each device creates a link-local address for itself based on the prefix FE80:: and the interface MAC address. This address is only valid on the local network. It then uses NDP to make sure that the address is unique.

Note

IGP routing protocols use the link-local address to form neighbor relationships. It is also the next-hop address that is installed in the routing table by IGPs.

NDP has several functions in IPv6, including the following:

- **Duplicate Address Discovery (DAD):** The host uses Neighbor Solicitation (NS) to send a message to its own address. No response means that the link-local address is unique.
- **Neighbor Discovery:** Similar to ARP, the host discovers the link-local address of neighbors using an NS message. This is ICMP type 135. Neighbors respond with an ICMP type 136 message.
- **Router Discovery:** IPv6 routers periodically send Router Advertisements (RAs) listing the network prefix. When a host comes online it immediately sends a Router Solicitation (RS) message, asking for prefix information, rather than waiting for the RA. This is sent to the All-routers multicast address.

To configure stateless autoconfiguration, use the interface command **ipv6 address autoconfig**. Acquiring an address involves the following steps:

1. The host creates a link-local address.
2. It sends an NS message to its link-local address out the interface.
3. If there is no reply, DAD declares the address unique.
4. If the host doesn't receive an RA, it sends an RS.
5. A router on the subnet sends an RA, listing its interface prefix.

6. The host uses that prefix and the interface MAC address to create its IPv6 address.

Use the command **show ipv6 interface** to verify your configuration. The following example shows this command and the resulting IPv6 address.

```
R4(config)# int fa 0/0
R4(config-if)# ipv6 address autoconfig
!
R4# show ipv6 int fa0/0
FastEthernet0/0 is up, line protocol is up
 IPv6 is enabled, link-local address is FE80::21D:A1FF:FE6C:D238
 No Virtual link-local address(es):
 Global unicast address(es):
  2001:8::21D:A1FF:FE6C:D238, subnet is 2001:8::/64 [EUI/CAL/PRE]
   valid lifetime 2591828 preferred lifetime 604628
```

Renumbering

IPv6 supports easy network renumbering. Note in the previous example that lifetimes are listed for the subnet address. When it is time to change the subnet, you can configure the router to advertise the old prefix with a short lifetime and a new prefix with a longer lifetime. You can even configure the router to expire a prefix at a certain date and time. The router sends out an RA with both prefixes and their lifetimes. Hosts then update their addresses. Anyone who has had to renumber a large range of IPv4 addresses can testify to how useful this feature is!

IPv6 Routing

Routing with IPv6 will seem very familiar to you. The same IGPs—RIP, EIGRP, and OSPF—are used as in IPv4; they have been adapted to carry IPv6 routes. BGP extensions allow it to do IPv6 routing. The same rules for metric and administrative distance apply. The commands are very similar too. The main difference in commands is that you need to specify that the command pertains to IPv6, since IPv4 is the default. One big configuration difference is that the **network** command is no longer used by IGPs to initiate routing. It is enabled at each interface instead. BGP does still use the **network** command to designate which networks to advertise.

Static Routing

Static routing with IPv6 works exactly like it does with version 4. Aside from understanding the address format, there are no differences. The syntax for the IPv6 static route command is

```
Router(config)# ipv6 route ipv6-prefix/prefix-length {ipv6-address |
  interface-type
interface-number [ipv6-address]} [administrative-distance]
[administrative-multicast-distance | unicast | multicast] [tag tag]
```

The following examples show the command in context as it might be applied. The first line shows a recursive static route that lists a next-hop address. The second line shows a directly connected static default route that lists an outbound interface. The third line shows a fully specified static route, which lists both the next-hop address and the outbound interface.

```
RouterA(config)# ipv6 route 2001:0:1:2::/64 2001:0:1:1::1
RouterA(config)# ipv6 route ::/0 serial1/0/0
RouterA(config)# ipv6 route 2001:0:1:2::/64 serial1/0/1
 2001:0:1:1::1
```

Verify your configuration with the command **show ipv6 route**.

RIPng for IPv6

RIP next generation (RIPng) is the IPv6 version of RIP and is defined in RFC 2080. Like RIPv2 for IPv4, RIPng is a distance vector routing protocol that uses a hop count for its metric, has a maximum hop count of 15, and uses split horizon. It uses UDP and still has an administrative distance of 120. RIPng also uses periodic multicast updates—every 30 seconds—to advertise routes. The multicast address is FF02::9. (RIP v2 uses IPv4 address 224.0.0.9.) The source address of RIPng updates is the link-local address of the outbound interface.

There are two important differences between the old RIP and the next-generation RIP. First, RIPng supports multiple concurrent processes, each identified by a process number. (This is similar to OSPFv2.) Second, RIPng is initialized in global configuration mode and then enabled on specific interfaces. There is no **network** command in RIPng.

The following example shows the syntax used to apply RIPng to a configuration. Notice that the syntax is similar to traditional RIP. You must first enable IPv6 routing. The global command to enable RIPng is optional; the router creates it automatically when the first interface is enabled for RIPng. You might need the command for additional configuration, such as disabling split horizon for a multipoint interface, as shown in the example.

```
Router(config)# ipv6 router rip process
Router(config-rtr)# no split-horizon
!
Router(config)# interface type number
Router(config-if)# ipv6 rip process enable
```

Like RIP for IPv4, troubleshoot RIPng by looking at the routing table (**show ipv6 route [rip]**), by reviewing the routing protocols (**show ipv6 protocols**), and by watching routing updates propagated between routers (**debug ipv6 rip**).

EIGRP for IPv6

EIGRP has been expanded to support IPv6, although you need to verify that your specific version of IOS is capable of doing this. EIGRP for IPv6 is based on the IPv4 version, and the two can run in tandem on the same router and on the same interfaces. EIGRP is still an advanced distance vector routing protocol that uses a complex metric. EIGRP still has a reliable update mechanism and uses DUAL to retain fall-back paths. Like EIGRP in IPv4, it sends multicast hellos every 5 seconds. (But the multicast address is now FF02::A.) Messages are exchanged using the interface link-local address as the source address. This leads to the possibility that two routers with interfaces on different subnets can now form an EIGRP adjacency.

Like RIPng, there is no more **network** command; EIGRP routing is enabled at each interface. You must assign a router ID in the format of a 32-bit IPv4 address. It does not need to be a routable address. One important thing to note is that the protocol starts off in the shutdown state. You must **no shut** it before routing will begin. Auto-summarization is disabled by default in IPv6 EIGRP.

The following example shows how to enable IPv6 EIGRP:

```
Router(config)# ipv6 unicast-routing
!
Router(config)# ipv6 router eigrp AS
Router(config-rtr)# router-id ipv4-address
Router(config-rtr)# no shut
!
Router(config)# interface type number
Router(config-if)# ipv6 eigrp AS
```

Like EIGRP for IPv4, troubleshoot by looking at the routing table (**show ipv6 route**), by reviewing the routing protocols (**show ipv6 protocols**), and by monitoring neighbors (**show ipv6 eigrp neighbors**).

IPv6 EIGRP can summarize routes at the interface, and the stub feature is also available, just as with the IPv4 version. The following example shows a sample configuration for IPv6 EIGRP, with both summarization and stub routing enabled. Notice that the routing protocol is enabled under each interface:

```
RouterA(config)# ipv6 router eigrp 1
RouterA(config-rtr)# router-id 10.255.255.1
RouterA(config-rtr)# stub connected summary
!
RouterA(config)# interface fastethernet0/0
RouterA(config-if)# description Local LAN
RouterA(config-if)# ipv6 address 2001:0:1:1::2/64
RouterA(config-if)# ipv6 eigrp 1
!
RouterA(config-if)# interface serial 1/0/1
RouterA(config-if)# description point-to-point line to Internet
RouterA(config-if)# ipv6 address 2001:0:1:5::2/64
RouterA(config-if)# ipv6 eigrp 1
RouterA(config-if)# ipv6 summary address eigrp 1 2001:0:1/24
```

OSPFv3

OSPFv3 was one of the first routing protocols available for IPv6 and because of its open-standard heritage, it is widely supported in IPv6. OSPFv3, which supports IPv6, is documented in RFC 2740. Like OSPFv2, it is a link-state routing protocol that uses the Dijkstra algorithm to select paths. Routers are organized into areas, with all areas touching area 0.

OSPFv3 routers use the same packet types as OSPFv2, form neighbors in the same way, flood and age LSAs identically, and support the same NBMA topologies and techniques such as NSSA and on-demand circuits. It can run concurrently with OSPFv2 because each version maintains its own databases and runs a separate SPF calculation.

OSPFv3 differs from its predecessors principally in its new address format. OSPFv3 advertises using multicast addresses FF02::5 and FF02::6 but uses its link-local address as the source address of its advertisements. This means that OSPF can form adjacencies with neighbor routers that are not on the same subnet. Multiple instances of OSPFv3 can run on each link. Authentication is no longer built in but relies on the underlying capabilities of IPv6.

OSPFv3 configuration is similar to RIPng and EIGRP. The routing process is created and routing properties are assigned to it. As with EIGRP, you must create a router ID in 32-bit dotted decimal format. The router ID is not automatically created in OSPFv3. Interfaces are associated with the OSPF process under interface configuration mode.

Assuming that **ipv6 unicast-routing** and interface IP addresses are already in place, the commands to implement basic OSPFv3 are shown in the following example.

```
Router(config)# ipv6 router ospf process-id
Router(config-rtr)# router-id 32bit-address
!
Router(config-rtr)# interface type number
Router(config-if)# ipv6 ospf process-id area area
```

As illustrated in the following example, route summarization is still configured under the OSPF routing process. Stub routing is also configured under the routing process, using the same commands as with OSPFv2. The default costs and interface priorities can be overridden at each interface. This example shows how these commands might look on an actual router.

```
RouterA(config)# ipv6 unicast-routing
!
RouterA(config)# ipv6 router ospf 1
RouterA(config-rtr)# router-id 10.255.255.1
RouterA(config-rtr)# area 1 range 2001:0:1::/80
RouterA(config-rtr)# area 1 stub no summary
!
RouterA(config-rtr)# interface fastethernet0/0
RouterA(config-if)# ipv6 address 2001:0:1:1::2/64
RouterA(config-if)# ipv6 ospf 1 area 1
RouterA(config-if)# ipv6 ospf cost 10
RouterA(config-if)# ipv6 ospf priority 20
!
RouterA(config-if)# interface serial 1/0/0
RouterA(config-if)# ipv6 address 2001:0:1:5::1/64
RouterA(config-if)# ipv6 ospf 1 area 0
```

Troubleshoot OSPFv3 just like OSPFv2. Start by looking at **show ipv6 route** to verify routes have been advertised. Assuming the route is in the routing table, test reachability using **ping ipv6**. You can also look at the OSPF setup using **show ipv6 ospf** *process* **interface**, **show ipv6 ospf**, or **show ipv6 ospf database**.

MP-BGP for IPv6

Multiprotocol BGP (RFC 2858) involves two new extensions to BGP4 that enable BGP to carry reachability information for other protocols, such as IPv6, multicast IPv4, and MPLS. The extensions enable NEXT_HOP to carry IPv6 addresses and NLRI (network layer reachability information) to an IPv6 prefix. An **address-family** command is added to the BGP configuration to enable this.

Router ID must be manually configured in an all-IPv6 implementation and is a 32-bit dotted decimal number. Unlike the IGPs, configuration is done under the BGP router configuration mode, not at the interface. Neighbors are configured under the global BGP configuration mode but must be activated under the IPv6 address family mode. Any policies or networks relevant to this mBGP extension are also configured under the address family.

The following example shows the BGP commands as they might be applied.

```
RouterA(config)# ipv6 unicast-routing
!
RouterA(config)# router bgp 65000
RouterA(config-rtr)# router-id 10.255.255.1
RouterA(config-rtr)# neighbor 2001:0:1:1:5::4 remote-as 65001
RouterA(config-rtr)# address-family ipv6 unicast
RouterA(config-rtr-af)# neighbor 2001:0:1:5::4 activate
RouterA(config-rtr-af)# network 2001:0:1::/48
```

To verify your BGP configuration, use the commands **show bgp ipv6 unicast summary** and **show ipv6 route bgp**.

RIPng Redistribution

You can run multiple instances of RIPng on the same router by giving them different process tags in the global RIP configuration. Be sure to use the correct tag when you enable RIP on each interface.

Note

The process tag is case-sensitive.

An interesting thing about RIPng is that the multiple instances exchange routing information with each other if they use the same multicast group and UDP port number. To keep the route information separate, you need to configure each instance to use a different port. Do this under the global

RIPng configuration mode for each process. You can keep the default multicast group:

```
R1(config)# ipv6 router rip Process1
R1(config-rtr)# port 1010 multicast-group ff02::9
!
R1(config)# ipv6 router rip Process2
R1(config-rtr)# port 1011 multicast-group ff02::9
```

Remember to do this on all routers in the RIP process. If you need to share routes between the two processes, you can control the redistribution by configuring it on the desired routers. You can further control it by using a route map to modify the redistribution. With a route map you can set the seed metric for specific routes, or filter routes that should not be redistributed, just as you can with IPv4 routing. The command to redistribute Process2 routes into Process1 would look like this:

```
R1(config)# ipv6 router rip Process1
R1(config-rtr)# redistribute rip Process2 route-map Filter
```

Redistribution between other IPv6 routing protocols use the same commands and follow most of the same rules as IPv4 routing protocols.

Integrating IPv4 and IPv6

There are several strategies for migrating from IPv4 to IPv6. Each of these strategies should be considered when organizations decide to make the move to IPv6 because each has positive points to aiding a smooth migration. It should also be said that there does not have to be a global decision on strategy—your organization might choose to run dual-stack in the United States, go completely to IPv6 in Japan, and use tunneling in Europe. The transition mechanisms include

- **Dual stack:** Running IPv6 and IPv4 concurrently on the same interface.
- **Tunneling:** Routers that straddle the IPv4 and IPv6 worlds encapsulate IPv6 traffic inside IPv4 packets.
- **Translation:** Using an extension of NAT, NAT protocol translation (NAT-PT), to translate between IPv4 and IPv6 addresses.

Tunneling IPv6 over IPv4

A tunnel serves as a virtual point-to-point link between IPv6 domains. It doesn't matter what the underlying IPv4 structure is if there is IP reachability between the tunnel endpoints. This exam covers five ways to tunnel IPv6 over IPv4:

- Manual Tunnels
- GRE Tunnels
- 6to4 Tunnels
- IPv4-Compatible IPv6 Tunnels
- Intra-Site Automatic Tunnel Addressing Protocol (ISATAP)

Manual Tunnels

When you manually create the tunnel, the source and destination IP addresses are IPv4 addresses because IPv4 is the transport protocol. You might want to use loopback addresses for increased stability. IPv6 addresses go on the tunnel interfaces because IPv6 is the passenger protocol. Because IPv6 considers the tunnel a point-to-point link, the address of each end of the tunnel is in the same subnet. Include the command **tunnel mode IPv6IP** in tunnel configuration mode to enable IPv6 over IP encapsulation.

To verify your configuration you can use the commands **debug tunnel** or **show interface tunnel** *int-number*.

GRE Tunnels

GRE is the default tunnel mode for Cisco routers. It provides more flexibility because it is protocol-agnostic. It can carry multiple protocols and can use multiple protocols for its transport, including IPv6 and routing protocols.

Configuring an IPv4 GRE tunnel to carry IPv6 traffic is the same as configuring a manual tunnel except you do not have to specify the tunnel mode because GRE is the default. You can allow a routing protocol on the tunnel interface, too. The process is the same as enabling it on a physical interface.

To configure a completely IPv6 GRE tunnel, use IPv6 interface addresses as the tunnel source and destination. Give the tunnel endpoints IPv6 addresses, too. You need a command to identify that the transport protocol is IPv6. That command, given in tunnel configuration mode, is **tunnel mode gre ipv6**.

6to4 Tunnels

This technique dynamically creates tunnels that IPv6 considers point-to-multipoint interfaces. You use the reserved prefix 2002::/16 in your IPv6 domain and then add the IPv4 address of the dual-stack router on the other side of the IPv4 domain as the next 32 bits of the network address. This means you need to translate that IP address into hexadecimal.

When IPv6 traffic arrives at an edge dual-stack router with a destination IPv6 prefix of 2002::/16, the router looks at the first 48 bits, derives the embedded IPv4 address from them, and uses it to determine the packet destination. The router then encapsulates the IPv6 packet in an IPv4 packet with the extracted IPv4 address as the packet destination.

Configure a tunnel as before, using IPv4 addresses as the source, but do not manually specify a destination. Give the tunnel an IPv6 address as previously described, with the tunnel destination embedded in its prefix. The tunnel mode command is **tunnel mode ipv6ip 6to4**.

Each router needs a route to its peer on the other side of the IPv4 network. The only current options for this are static routes and BGP.

IPv4-Compatible IPv6 Tunnels

This type of tunnel has been deprecated. It encodes the IPv4 address of the tunnel source in the lowest 32 bits of the IPv6 tunnel address and then pads the rest of the bits with zeros. It uses the tunnel mode command **tunnel mode ipv6ip auto-tunnel**.

ISATAP Tunnels

ISATAP tunnels are similar to the other two tunnels techniques in that an IPv4 address is encoded into the IPv6 address. It is meant to be used within a site, between hosts and routers, although it can be used between sites.

The tunnel source address is an IPv4 address. Do not specify a tunnel destination. The IPv6 address of the tunnel itself combines the network prefix, 0000:5EFE, and the 32-bit IPv4 tunnel source address. The IPv4 address is encoded into the least significant 32 bits of the address. You can use any network prefix. The tunnel interface link-local address still starts with FE80 and then uses 0000:5EFE plus the encoded IPv4 address.

For instance, the link-local address of a tunnel that uses 10.8.8.8 as its source is

FE80::5EFE:A08:808

The unicast IPv6 address of that same tunnel interface, assuming that prefix 2001:1:2:3/64 was assigned to the interface, is

2001:1:2:3:0:5EFE:A08:808

ISATAP tunnels do not support multicast. A route is needed to the tunnel destination if it is in a different subnet; this can be either a static route or a BGP route.

Using Address Translation

Instead of replacing IPv4, there are several ways to coordinate the functioning of IPv4 and v6 concurrently. NAT-Protocol Translation is an example of this coexistence strategy. NAT-PT does bidirectional translation between IPv4 and IPv6 addresses. Use it when hosts using IPv4 need to establish a session with hosts using IPv6, and vice versa. If hosts communicate using DNS names, a DNS Application Layer Gateway (DNS ALG) can resolve names to both IPv4 and IPv6 addresses.

To enable NAT-PT on a router, use the command **ipv6 nat** on each interface in which traffic needs to be translated. You must also configure at least one NAT-PT prefix. This is used to decide which traffic to NAT; only traffic matching the prefix will be translated. This is configured either at the global configuration mode (to apply to the entire router) or at the interface configuration mode (to apply only to traffic on that interface.) The command is **ipv6 nat prefix** *prefix/prefix-length.*

Static NAT-PT

You can use either static or dynamic mapping of addresses. To configure static translation of an IPv6 address to an IPv4 address, use the global command:

```
ipv6 nat v6v4 source ipv6-address ipv4-address
```

To configure static mapping of an IPv4 address to an IPv6 address, use the global command:

```
ipv6 nat v4v6 source ipv4-address ipv6-address
```

Dynamic NAT-PT

Dynamic mapping draws from a pool of addresses to temporarily assign to hosts. You need to create a pool of addresses and then configure NAT-PT to use that pool. You can optionally control the traffic to be mapped by using an access list or route map. To create the pool and enable NAT-PT for IPv4 to IPv6 translation, use the global commands:

ipv6 nat v4v6 pool *name start-ipv6 end-ipv6* **prefix-length** *prefix-length*

ipv6 nat v4v6 source list {*access-list-number* | *name*} **pool** *name*

To create the pool and enable NAT-PT for IPv6 to IPv4 translation, use the global commands:

ipv6 nat v6v4 pool *name start-ipv4 end-ipv4* **prefix-length** *prefix-length*

ipv6 nat v6v4 source {**list** *access-list-name* | **route-map** *map-name*} **pool** *name*

Verify NAT-PT operation with the commands **show ip nat translations**, **show ip nat statistics**, **show ipv6 nat translations**, and **show ipv6 nat statistics**.

IPv6 Link Types

IPv6 recognizes three types of links:

- Point-to-point
- Point-to-multipoint
- Multiaccess

Point-to-Point Links

Recall that an IPv6 interface uses its MAC address to create its link-local address. A serial link has no MAC address associated with it, so it uses one from an Ethernet interface. You can manually configure the link-local address to make it more recognizable. Be sure to begin the IPv6 address with the link-local prefix FE80.

Point-to-point links do not necessarily need global unicast addresses. The routers can communicate with only link-local addresses, but you could not reach those interfaces from off the network because the link-local is not a routable address.

Point-to-Multipoint Links

For point-to-multipoint links, such as Frame Relay, you must map the destination IPv6 address to the correct DLCI, just as with IPv4. The difference is that with IPv6 you must also map the link-local address to the DLCI because it is used as the next hop for routing. So for each DLCI, you must have at least two mappings: the remote router's IPv6 global unicast address and the remote router's link-local address. The map command is

```
frame relay map ipv6 destination-address out dlci dlci-number broad-
  cast
```

In a hub-and-spoke topology, the hub must be configured for IPv6 unicast routing for the spokes to communicate with each other.

Multiaccess Links

Devices on multiaccess links, such as Ethernet, build a table mapping destination Layer 3 addresses to Layer 2 addresses, whether you use IPv4 or IPv6. IPv4 uses a separate protocol, ARP, to do this. In IPv6 the process is built into the IPv6 protocol with the Neighbor Discovery process. It uses ICMPv6.

An IPv6 device sends a Neighbor Solicitation (NS) multicast with a prefix of FE02. The neighbor responds with a Neighbor Advertisement (NA) message listing its MAC address. As with ARP, these mappings have a set lifetime (called the *reachable time*), so an NS can also be sent periodically to verify that a neighbor is still reachable.

To add a static entry to the Neighbor Discovery table, use the command **ipv6 neighbor** *ipv6-address interface-type interface-number hardware-address*. A static address does not age out of the table.

Display the mappings with the **show ipv6 neighbors** command.

APPENDIX A

Understanding IPsec

You are not required to have detailed knowledge of IPsec for the ROUTE exam. This Appendix is intended to help solidify your understanding of the technology. It can also serve as a command reference if you need to actually configure IPsec.

IPSecurity, or IPsec, is a set of rules for securing data communications across a public, untrusted network such as the Internet. It provides

- Data confidentiality by encrypting portions of a packet
- Data integrity by ensuring the packet has not been altered in transit
- Data source authentication to ensure the data originated with a trusted source
- Antireplay protection to ensure that packets are not copied and sent

IPsec standards do not specify exactly how packets should be encrypted or authenticated; it relies on other protocols to accomplish those functions. For encryption it can use Data Encryption Standard (DES), Triple Data Encryption Standard (3DES), and Advanced Encryption Standard (AES). For authentication it can use Hash-based Message Authentication Codes (HMAC). An HMAC combines a hash function such as Message Digest 5 (MD5), and Secure Hash Algorithm 1 (SHA-1) with a shared secret key. MD5 uses a 128-bit hash, whereas SHA-1 uses a 160-bit hash; however, IPsec uses only 96 bits of the SHA-1 hash.

IPsec Headers

IPsec defines two types of headers: Authentication Header and Encapsulating Security Payload.

Authentication Header

Authentication Header (AH) is IP protocol number 51. It authenticates the packet, including the IP header, but does not encrypt the packet payload. AH works by creating an MD5 or SHA-1 hash from the IP header (except any changeable fields such as Time to Live) and the packet payload. It sends this hash in an AH header after the Layer 3 IP header. The receiving host also creates a hash value from the IP header and packet payload, and compares the two hashes. If they match, the packet was unchanged during transit. A shared key creates the hashes, so a match also serves to authenticate the source of the packet, which is rarely used without ESP.

Encapsulating Security Payload

Encapsulating Security (ESP), IP protocol number 50, encrypts packet payloads and can optionally authenticate and do integrity checks by using it with AH. It adds a header and a trailer to the packet. When used with AH, the packet is encrypted first and then put through the hash mechanism.

IPsec Modes

IPsec can operate in either transport mode or tunnel mode. The headers differ based on the mode used:

- Transport Mode IPsec uses the original IP header. The data payload can be encrypted, and the packet can be authenticated from the ESP header back. Transport mode is often used with Generic Routing Encapsulation (GRE) tunnels because GRE hides the original IP address.
- Tunnel Mode IPsec replaces the original IP header with a tunnel header. The ESP header is placed after the new header, before the original one. The original IP header can be encrypted along with the data payload, and the packet can be authenticated from the ESP header back. Tunnel mode adds approximately 20 bytes to the packet.

Figure A-1 shows the packet headers in the two IPsec modes.

Figure A-1 Transport Mode Versus Tunnel Mode IPsec

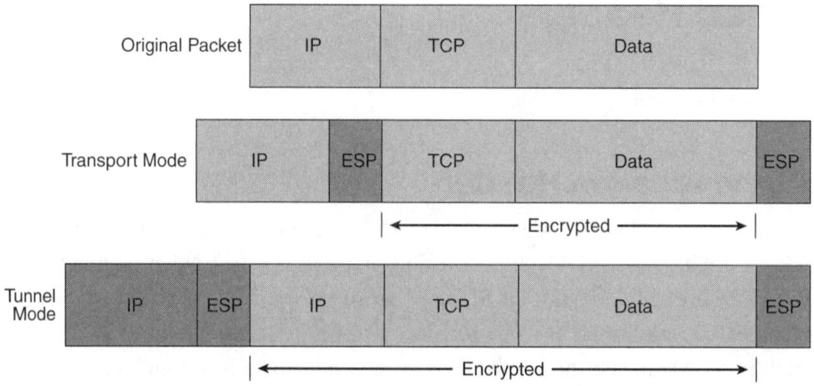

Tunnel mode ESP can cause problems when used with Network Address Translation (NAT). The original TCP or UDP header is encrypted and hidden, so there are no Layer 4 port numbers for NAT to use. *NAT Traversal* detects the existence of a NAT device and adds a UDP header after the tunnel IP header. NAT can then use the port number in that UDP header.

Authentication Methods

Several authentication methods are supported with IPsec VPNs:

- Username and password
- A one-time password
- Biometric features, such as fingerprint
- Preshared key values
- Digital certificates

Encryption Methods

IPsec encryption uses key values to encrypt and decrypt data. Keys can be either symmetric or asymmetric. Symmetric keys use the same value to both encrypt and decrypt the data, which include DES, 3DES, and AES. Asymmetric keys use one value to encrypt the data and another one to decrypt it. Diffie-Hellman and RSA use asymmetric keys.

NOTE

RSA is not an acronym; it is the initials of the last names of the algorithm's inventors: Ron Rivest, Adi Shamir, and Len Adleman.

Symmetric Key Algorithms

DES uses a 56-bit key and can be broken fairly easily. It is a block cipher that encrypts 64-bit blocks of data at a time.

3DES is also a block cipher, but it encrypts each block, decrypts it, and then encrypts it again. A 56-bit key is used each time, thus equaling a key length of 168 bits. It is more secure than DES but also requires more processing power.

AES is a stronger block cipher encryption method than DES or 3DES. It uses a 128-bit data block and a key length of 128 bits, 192 bits, or 256 bits. AES has been approved for use with government classified data.

Asymmetric Key Algorithm

RSA uses asymmetric keys and can be used for signing messages and encrypting them. A public key encrypts or signs the data. It can be decrypted only with a private key held by the receiver. RSA is slower than symmetrical key algorithms but more secure if a large enough key is used. A key length of 2048 bits is recommended.

Diffie-Hellman Key Exchange

The Diffie-Hellman protocol solves the problem of exchanging keys over an insecure network. Each device creates a public key and a private key. They exchange their public keys in the open, unencrypted. They each then use the other device's public key and their own private key to generate a shared secret key that each can use.

Key Management

The public key infrastructure (PKI) manages encryption and identity information such as public keys and certificates. It consists of the following components:

- Peer devices that need to communicate securely.
- Digital certificates that validate the peer's identity and transmit their public key.
- Certificate authorities (CA), also known as trustpoints, that grant, manage, and revoke certificates. This can be a third-party CA or an internal one. Cisco has an IOS Certificate Server.
- Optional Registration authorities (RA) that handle certificate enrollment requests.
- A way to distribute certificate revocation lists (CRL), such as HTTP or Lightweight Directory Access Protocol (LDAP).

PKI credentials, such as RSA keys and digital certificates, can be stored in a router's NVRAM. The can also be stored in USB eTokens on routers that support them.

Establishing an IPsec VPN

When IPsec establishes a VPN between two peer hosts, it sets up a security association (SA) between them. SAs are unidirectional, so each bidirectional data session requires two. The Internet Security Association and Key Management Protocol (ISAKMP) defines how SAs are created and deleted. Following are five the basic steps:

1. **Interesting traffic arrives at the router:** "Interesting" traffic is that which should be sent over the VPN. This is specified by a crypto access list. Any traffic not identified as "interesting" is sent in the clear, unprotected.
2. **Internet Key Exchange (IKE) Phase One:** Negotiates the algorithms and hashes to use, authenticates the peers, and sets up an ISAKMP SA. This has two modes: Main and Aggressive. Main mode uses three exchanges during Phase One. Aggressive mode sends all the information in one exchange. The proposed settings are contained in *Transform Sets* that list the proposed encryption algorithm, authentication algorithm, key length, and mode. Multiple transform sets can be specified, but both peers must have at least one matching transform set or the session is torn down.
3. **IKE Phase Two:** Uses the secure communication channel created in Phase One to set up the SAs for ESP and AH, negotiating the SA parameters and settings to be used to protect the data transmitted. This periodically renegotiates the SAs. SAs have lifetimes that can be measured in either the amount of data transferred or length of time. Might do an additional Diffie-Hellman key exchange during Phase Two.
4. **Data is transferred along the VPN between the two peers:** It is encrypted by one peer and decrypted by the other, according to the transform sets negotiated.
5. **Tunnel termination:** The IPsec session drops because of either direct termination or time out.

Configuring a Site-to-Site VPN Using IOS

Following are six steps to configuring a site-to-site IPsec VPN using Cisco IOS commands:

1. Configure the ISAKMP policy.
2. Configure the IPsec transform set or sets.

3. Configure a crypto access control list (ACL).
4. Configure a crypto map.
5. Apply the crypto map to the outgoing interface.
6. Optionally configure and apply an ACL that permits only IPsec or IKE traffic.

Configuring an ISAKMP Policy

To configure an ISAKMP policy, first create the policy and then give the parameters. These parameters might include such things as type of encryption, type of hash, type of authentication, SA lifetime, and Diffie-Hellman group. The following example shows an ISAKMP policy configuration, along with the options available with each parameter. Options can vary based on IOS version.

```
IPSEC_RTR(config)# crypto isakmp policy ?
  <1-10000>  Priority of protection suite
IPSEC_RTR(config)# crypto isakmp policy 1
!
IPSEC_RTR(config-isakmp)# encryption ?
  3des   Three key triple DES
  aes    AES - Advanced Encryption Standard.
  des    DES - Data Encryption Standard (56 bit keys).
IPSEC_RTR(config-isakmp)# encryption 3des
!
IPSEC_RTR(config-isakmp)# hash ?
  md5  Message Digest 5
  sha  Secure Hash Standard
IPSEC_RTR(config-isakmp)# hash sha
!
IPSEC_RTR(config-isakmp)# authentication ?
  pre-share  Pre-Shared Key
  rsa-encr   Rivest-Shamir-Adleman Encryption
  rsa-sig    Rivest-Shamir-Adleman Signature
IPSEC_RTR(config-isakmp)# authentication pre-share
!
IPSEC_RTR(config-isakmp)# group ?
  1  Diffie-Hellman group 1
  2  Diffie-Hellman group 2
  5  Diffie-Hellman group 5
IPSEC_RTR(config-isakmp)# group 2

IPSEC_RTR(config-isakmp)# lifetime ?
  <60-86400>  lifetime in seconds
IPSEC_RTR(config-isakmp)# lifetime 300
```

Configuring an IPsec Transform Set

An IPsec transform set defines how VPN data will be protected. It specifies the IPsec protocols that will be used. You can specify up to four transforms and the algorithm to use with them. You can also configure either tunnel or transport mode. (Tunnel is default.) The transforms include

- AH with either MD5 or SHA-1
- ESP encryption using DES, 3DES, AES, or others
- ESP authentication using MD5 or SHA-1
- Compression using the Lempel-Ziv-Stac (LZS) algorithm

The following example shows a transform set with ESP encryption and authentication. Note that these commands are all given as part of the same command.

```
IPSEC_RTR# conf t
Enter configuration commands, one per line.  End with CNTL/Z.
IPSEC_RTR(config)# crypto ipsec transform-set TRANSFORM1 esp-aes 192
 esp-md5-hmac
```

Configuring a Crypto ACL

You use a crypto ACL to identify traffic that should be protected by the IPsec VPN, in particular the interesting traffic that brings up the tunnel. Any traffic permitted in the ACL is sent over the VPN. Traffic denied by the ACL is not dropped—it is simply sent normally.

The following example shows a crypto ACL that permits traffic from two internal networks—172.16.1.0 and 172.16.4.0—if it is bound to the server network of 10.6.3.0.

NOTE

When configuring the crypto ACL on the router at the other end of the tunnel, be sure to reverse the source and destination IP addresses.

```
IPSEC_RTR(config)  access-list 172 permit ip 172.16.1.0 0.0.0.255
 10.6.3.0 0.0.0.255
IPSEC_RTR(config)  access-list 172 permit ip 172.16.4.0 0.0.0.255
 10.6.3.0 0.0.0.255
```

Configuring a Crypto Map

A crypto map pulls together the transform sets and crypto ACLs, and associates them with a remote peer. You can use a sequence number when configuring a crypto map. Multiple crypto maps with the same name but different sequence numbers form a crypto map set. Traffic is evaluated against each crypto map depending on its sequence number to see if it should be protected. This permits more complex and granular traffic filtering.

The following example shows a crypto map that links the transform set and ACL configured in previous examples.

```
IPSEC_RTR(config)# crypto map TO_SERVERS 10 ipsec-isakmp
% NOTE: This new crypto map will remain disabled until a peer
        and a valid access list have been configured.
IPSEC_RTR(config-crypto-map)# set peer 10.1.1.1
IPSEC_RTR(config-crypto-map)# match address 172
IPSEC_RTR(config-crypto-map)# set transform-set TRANSFORM1
```

Applying the Crypto Map to an Interface

After the crypto map is configured, it must be applied to an interface for it to take effect. It is applied at the *outgoing* interface—the one that VPN traffic uses to reach the other end of the VPN. You might need to use a static route or otherwise adjust your routing to force traffic bound for the VPN destination networks to use the correct outgoing interface.

The following example shows the crypto map TO_SERVERS applied to interface serial 0/0/0. Note that the router replies with a message that ISAKMP is now enabled:

```
IPSEC_RTR(config)# int s0/0/0
IPSEC_RTR(config-if)# crypto map TO_SERVERS
IPSEC_RTR(config-if)#
01:19:16: %CRYPTO-6-ISAKMP_ON_OFF: ISAKMP is ON
```

Configuring an Optional Interface Access List

You might want to have an interface ACL on the VPN interface. Typically you would permit only IPsec-related traffic, and perhaps routing protocol traffic, in and out that interface. Keep in mind the following port numbers when configuring the ACL:

- ESP is IP protocol 50.
- AH is IP protocol 51.

- IKE uses UDP port 500.
- NAT traversal uses UDP port 4500.

The source and destination addresses should be the IP addresses of the outgoing VPN interfaces. The following example shows an ACL that permits IPsec traffic between two hosts.

```
IPSEC_RTR(config)# access-list 101 permit ahp host 10.1.1.2 host
  10.1.1.1
IPSEC_RTR(config)# access-list 101 permit esp host 10.1.1.2 host
  10.1.1.1
IPSEC_RTR(config)# access-list 101 permit udp host 10.1.1.2 eq isakmp
  host 10.1.1.1
IPSEC_RTR(config)# access-list 101 permit udp host 10.1.1.2 host
  10.1.1.1 eq isakmp
!
IPSEC_RTR(config)# interface s 0/0/0
IPSEC_RTR(config-if)# ip address 10.1.1.2 255.255.255.252
IPSEC_RTR(config-if)# ip access-group 101 out
```

Monitoring and Troubleshooting IPsec VPNs

Some useful IOS commands for monitoring your IPsec VPNs include

- **show crypto isakmp sa:** Shows all the IKE security associations currently active on the router. Look for a status of QM_IDLE to verify that the SA is active.
- **show crypto ipsec sa:** Shows the parameters used by each SA and traffic flow. Look for the count of packets being encrypted and decrypted to verify the VPNs operation.

To troubleshoot VPN problems, first verify IP connectivity. If that exists, review your configuration one more time. If the configuration looks correct on both peers, you can view detailed information about the IKE negotiations by using the command **debug crypto isakmp**.

Using GRE with IPsec

GRE is a tunneling protocol that can support multiple Layer 3 protocols, such as IP, IPX, and AppleTalk. It also enables the use of multicast routing protocols across the tunnel. It adds a 20-byte IP header and a 4-byte GRE header, hiding the existing packet headers. The GRE header contains a flag field and a protocol type field to identify the Layer 3 protocol being transported. It might optionally contain a tunnel checksum, tunnel key, and tunnel sequence number. GRE does not encrypt traffic or use any strong security measures to protect the traffic.

GRE can be used along with IPsec to provide data source authentication and data confidentiality and ensure data integrity. GRE over IPsec tunnels are typically configured in a hub-and-spoke topology over an untrusted WAN to minimize the number of tunnels that each router must maintain.

Figure A-2 shows how the GRE and IPsec headers work together.

Figure A-2 GRE over IPsec Headers

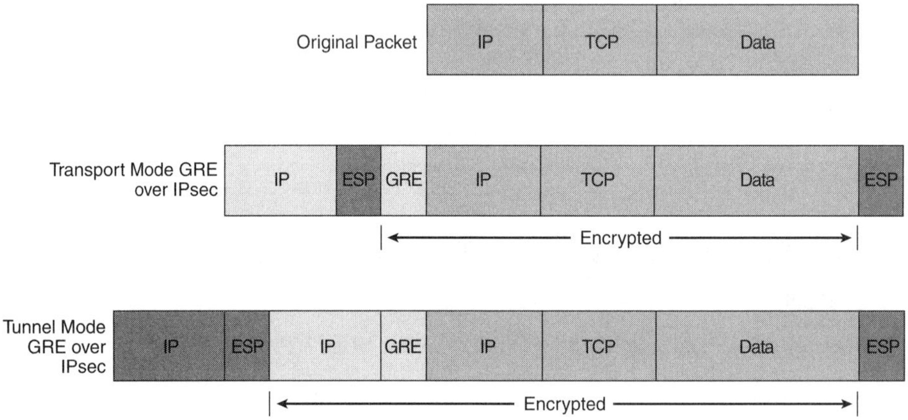

Configuring a GRE Tunnel Using IOS

To configure GRE using IOS commands, you must first configure a logical tunnel interface. GRE commands are then given under that interface. You must specify a source and destination for the tunnel; the source is a local outgoing interface. You might also give the tunnel interface an IP address and specify the tunnel mode. GRE is the default mode.

The following example shows a tunnel interface configured for GRE. The **mode** command is shown only as a reference because it is the default; it would not normally appear in the configuration.

```
interface Tunnel1
 ip address 172.16.5.2 255.255.255.0
 tunnel source Serial0/0/0
 tunnel destination 10.1.1.1
 tunnel mode gre ip
```

APPENDIX B

IPv6 Header Format

Although this specific material is not tested on the ROUTE exam, it might help you gain a better understanding of the structure of IPv6.

The IPv6 header is similar to the IPv4 header. The largest changes have to do with the larger addresses, aligning fields to 64-bit boundaries, and moving fragmentation to an extension header.

The fields are

- **Version:** 6.
- **Priority:** Similar to DSCP in version 4, this 8-bit field describes relative priority.
- **Flow:** 20-bit flow label enables tagging in a manner similar to MPLS.
- **Length:** The length of the data in the packet.
- **Next Header:** Indicates how the bits after the IP header should be interpreted. Can indicate TCP or UDP, or it can show an extension header.
- **Hop Limit:** Similar to TTL.
- **Source and Destination:** IPv6 addresses.

Zero or more extension headers can follow, including

- **Hop-by-hop options:** Options for intermediate devices.
- **Destination options:** Options for the end node.
- **Source routing:** Specifies "way stations" that the route must include.
- **Fragmentation:** Divides packets.
- **Authentication:** Attests to source. Replaces the AH header from IPsec.
- **Encryption:** Replaces the IPsec ESP header.

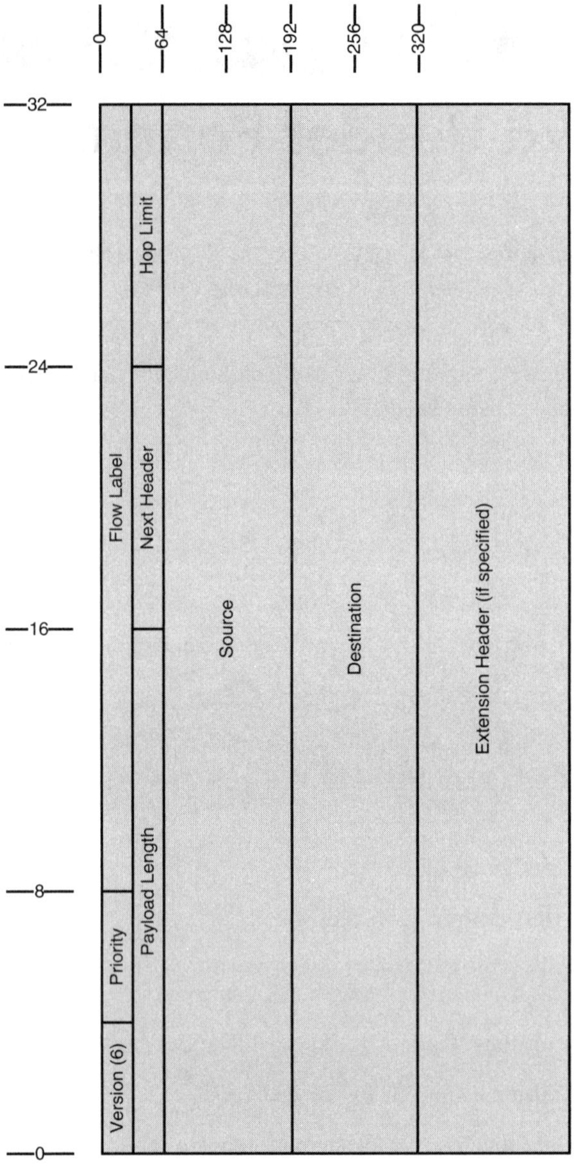

Figure B-1 IPv6 Header